OUR BEST
200 MAGNETS
IN PLASTIC CANVAS

Everybody loves magnets! These fun little projects are easy to finish in a jiffy, whether you're celebrating a holiday, showing your interest in a hobby, or just adding color to your message center. To bring you a diverse collection of magnet designs, we decided to pull together all the best magnets that have ever been published in Leisure Arts magazines, leaflets, and calendars. You'll find something for everyone and every taste! And don't forget how versatile magnets are — you can also use these designs to make plant pokes, decorate wreaths, or dress up all kinds of items. Our easy-to-read color charts and step-by-step instructions ensure that every minute of your stitching will be a pleasure, so let's get started! With every turn of the page, you'll see why we're proud to call this Our Best 200 Magnets.

LEISURE ARTS, INC.
Little Rock, Arkansas

OUR BEST 200 MAGNETS IN PLASTIC CANVAS

EDITORIAL STAFF

Vice President and Editor-in-Chief: Anne Van Wagner Childs
Executive Director: Sandra Graham Case
Executive Editor: Susan Frantz Wiles
Publications Director: Carla Bentley
Creative Art Director: Gloria Bearden
Production Art Director: Melinda Stout

PRODUCTION
Managing Editors: Teal Lee Elliott and Lisa Truxton Curton
Senior Editor: Donna Brown Hill
Project Coordinators: Phyllis Miller Boorsma, Michelle Sass Goodrich, Catherine Hubmann, Susan McManus Johnson, and Rhonda Goerke Lombardo
Project Assistant: Kandi Brock Ashford

EDITORIAL
Associate Editor: Linda L. Trimble
Senior Editorial Writer: Laura Lee Weland
Editorial Associates: Tammi Williamson Bradley, Terri Leming Davidson, and Robyn Sheffield-Edwards

ART
Crafts Art Director: Rhonda Hodge Shelby
Senior Production Artist: Katie Murphy
Production Artists: Roberta Aulwes, Hubrith E. Esters, Jonathan M. Flaxman, Gregory A. Needels, Susan Gray Vandiver, Dana Vaughn, Mary Ellen Wilhelm, and Karen L. Wilson
Photography Stylists: Laura Bushmiaer, Aurora Huston, Laura McCabe, and Emily Minnick

BUSINESS STAFF

Publisher: Bruce Akin
Vice President, Finance: Tom Siebenmorgen
Vice President, Retail Sales: Thomas L. Carlisle
Retail Sales Director: Richard Tignor
Vice President, Retail Marketing: Pam Stebbins
Retail Customer Services Director: Margaret Sweetin

Marketing Manager: Russ Barnett
Executive Director of Marketing and Circulation: Guy A. Crossley
Circulation Manager: Byron L. Taylor
Print Production Manager: Laura Lockhart
Print Production Coordinator: Nancy Reddick Lister

Library of Congress Catalog Number 95-76323
International Standard Book Number 0-942237-80-3

TABLE OF CONTENTS

3

PRIMARILY MONSTERS

These zany monster magnets will provide hours of fun for you and the kids. With our assortment of mix-and-match body parts and your choice of bright primary colors, you can create an endless array of kooky creatures.

PRIMARILY MONSTERS

Approx Size: 5¹⁄₂"w x 4³⁄₄"h each

Supplies For One Magnet: Worsted weight yarn (refer to photo), one 10¹⁄₂" x 13¹⁄₂" sheet of 7 mesh plastic canvas, #16 tapestry needle, two 12mm moving eyes, two 6" long black chenille stems, magnetic strip, and clear-drying craft glue

Stitches Used: Cross Stitch, Fringe, Gobelin Stitch, Overcast Stitch, Tent Stitch, and Turkey Loop

Instructions: Refer to photo for yarn colors. Follow charts and use required stitches to work desired Monster Magnet pieces. Refer to photo and use yarn color of Nose to join Nose to Body. Glue one Hand to each end of one chenille stem. Glue one Foot to each end of remaining chenille stem. Refer to photo to glue chenille stems to wrong side of Body. Refer to photo to add bows, Fringe hair, eyebrows, and mustache as desired. Refer to photo to glue moving eyes to Body. Glue magnetic strip to wrong side of stitched piece.

Designs by Jack Peatman for LuvLee.

▨ yarn

⊙ yarn Turkey Loop

Nose A
(6 x 10 threads)

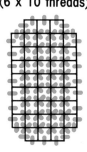

Nose B
(7 x 8 threads)

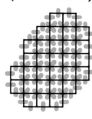

Nose C
(7 x 8 threads)

Nose D
(6 x 7 threads)

Nose E
(7 x 7 threads)

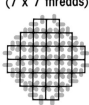

Oval Body (19 x 23 threads)

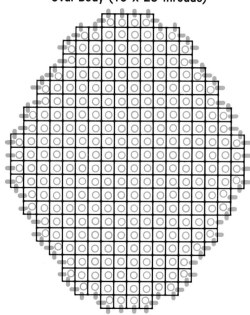

Round Body (19 x 19 threads)

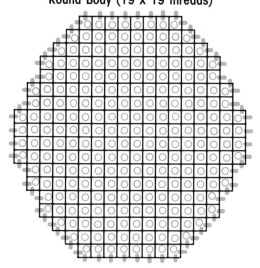

Right Hand
(7 x 8 threads)

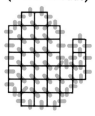

Left Hand
(7 x 8 threads)

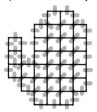

Right Foot
(10 x 5 threads)

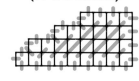

Left Foot
(10 x 5 threads)

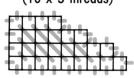

VALENTINE FRAMES

Heart Size: 3¼"w x 3¼"h

Love Size: 7½"w x 2¾"h

Supplies: Worsted weight yarn (refer to color key), one 10½" x 13½" sheet of 7 mesh plastic canvas, #16 tapestry needle, nylon line, magnetic strip, and clear-drying craft glue

Stitches Used: Backstitch, Overcast Stitch, and Tent Stitch

Instructions: Follow chart and use required stitches to work Front. Glue magnetic strip to Back. **For Heart Only:** Use pink to join Back to Front along unworked edges of Front. **For Love Only:** Place Back on wrong side of Front, covering opening. Use nylon line to tack Back to wrong side of Front along bottom and side edges of Back.

Heart design by Esther Metras.
Love design by Rose D. Munz.

Heart Front (22 x 21 threads)

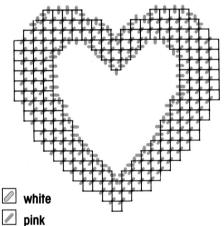

✎ white
✎ pink

Heart Back (22 x 19 threads)

Love Back (10 x 10 threads)

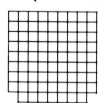

Love Front (51 x 18 threads)

CONVERSATION HEARTS

It's fun to leave love notes, and we've made it easy, too! Our conversation heart magnets were inspired by the popular candy hearts that sweetened our childhood days. They're perfect for sharing on Valentine's Day, but you'll want to give them as little surprises all year long.

Conversation Hearts (14 x 13 threads each)

CONVERSATION HEARTS

Size: 2¼"w x 2"h each

Supplies: Worsted weight yarn (refer to color key), red 6-strand embroidery floss, one 10½" x 13½" sheet of 7 mesh plastic canvas, #16 tapestry needle, magnetic strip, and clear-drying craft glue

Stitches Used: Backstitch, French Knot, Overcast Stitch, and Tent Stitch

Instructions: Follow chart and use required stitches to work Magnet. Glue magnetic strip to wrong side of stitched piece.

▱	white
▱	yellow
▰	peach
▱	pink
▰	lavender
▱	green
▱	red embroidery floss
⊙	red embroidery floss Fr. Knot

7

FOUR SEASONS MINI CALENDAR

Decorated with scenes of the four seasons, our mini calendar holder is a timeless accessory. The quick-and-easy design, which holds a wallet-size calendar, also can be made into a magnet.

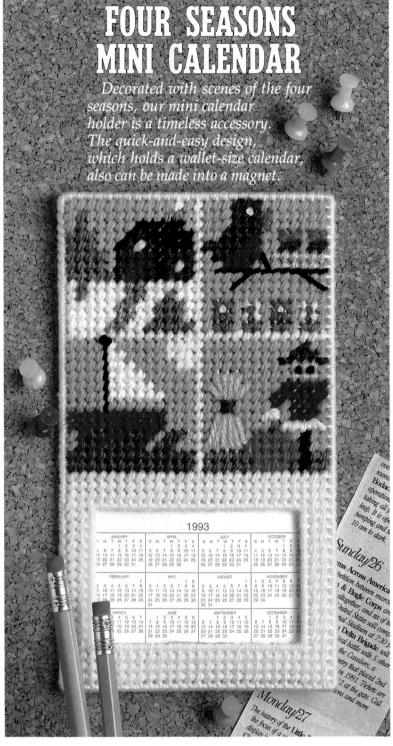

FOUR SEASONS MINI CALENDAR

Size: 4¼"w x 7½"h

(**Note:** Holds a 3½"w x 2¼"h pocket calendar.)

Supplies: Worsted weight yarn (refer to color key), one 10½" x 13½" sheet of 7 mesh plastic canvas, #16 tapestry needle, nylon line, magnetic strip, and clear-drying craft glue

Stitches Used: Backstitch, French Knot, Gobelin Stitch, Overcast Stitch, and Tent Stitch

Instructions: Follow chart and use required stitches to work Front. For Back, cut a piece of plastic canvas 27 x 19 threads. (**Note:** Back is not worked.) Place Back on wrong side of Front, covering opening. Use nylon line to join Back to Front along side and bottom edges of Back. Slide calendar in between Front and Back. Glue magnetic strip to wrong side of Magnet.

Design by Mary T. Cosgrove.

Front (29 x 50 threads)

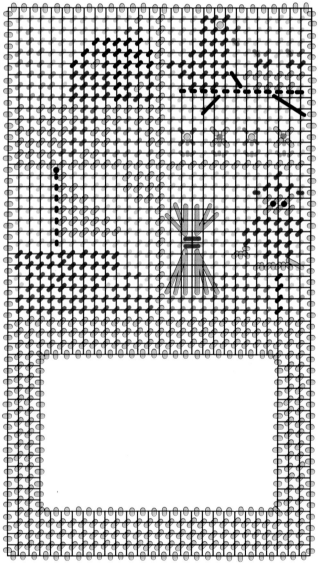

▨ white	▨ beige	
▨ yellow	▨ brown	
▨ orange	▨ grey	
▨ red	▨ black	
▨ green	◉ yellow 1 ply Fr. Knot	
▨ lt blue	◉ orange 1 ply Fr. Knot	
▨ blue	● black 1 ply Fr. Knot	

BRIGHT BALLOON FRAMES

These bright balloon magnets make ideal frames for your favorite photos. Each uplifting accent is also perfect for posting your memos, shopping lists, and other reminders.

BRIGHT BALLOON FRAMES

Size: 3 1/8"w x 4"h

(**Note:** Holds a 1 5/8"w x 2 1/4"h photograph.)

Supplies: Worsted weight yarn or Needloft® Plastic Canvas Yarn (refer to photo for colors), one 10 1/2" x 13 1/2" sheet of 7 mesh plastic canvas, #16 tapestry needle, 16" of white string, magnetic strip, and clear-drying craft glue

Stitches Used: Overcast Stitch and Tent Stitch

Instructions: Follow chart and use required stitches to work Front. Match △'s and ✗'s to join Back to Front along unworked edges of Front. Refer to photo to tie string around balloon. Glue magnetic strip to Back. Slide photograph between Front and Back.

Design by Elfie Moore.

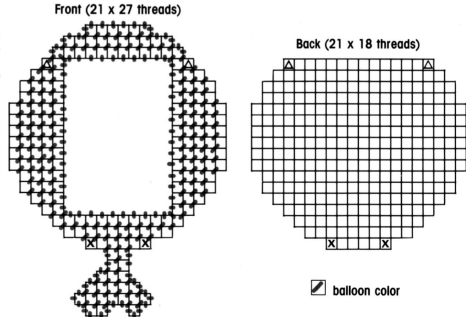

Front (21 x 27 threads)

Back (21 x 18 threads)

✏ balloon color

TROPICAL APPEAL

These colorful tropical bird magnets are sure to attract lots of attention to your notes and reminders. You can also use them as plant pokes or to add a sunny touch to a summertime wreath or other projects.

TROPICAL APPEAL

Approx Size: 3³/₄"w x 4¹/₄"h each

Supplies For Entire Set: Worsted weight yarn (refer to color keys), one 10¹/₂" x 13¹/₂" sheet of 7 mesh plastic canvas, #16 tapestry needle, five 6mm moving eyes, magnetic strip, and clear-drying craft glue

Stitches Used For Entire Set: Backstitch, Gobelin Stitch, Overcast Stitch, and Tent Stitch

Instructions: Follow chart and use required stitches to work Magnet. Refer to photo to glue moving eye to Magnet. Glue magnetic strip to wrong side of Magnet.

Designs by Virginia Hockenbury.

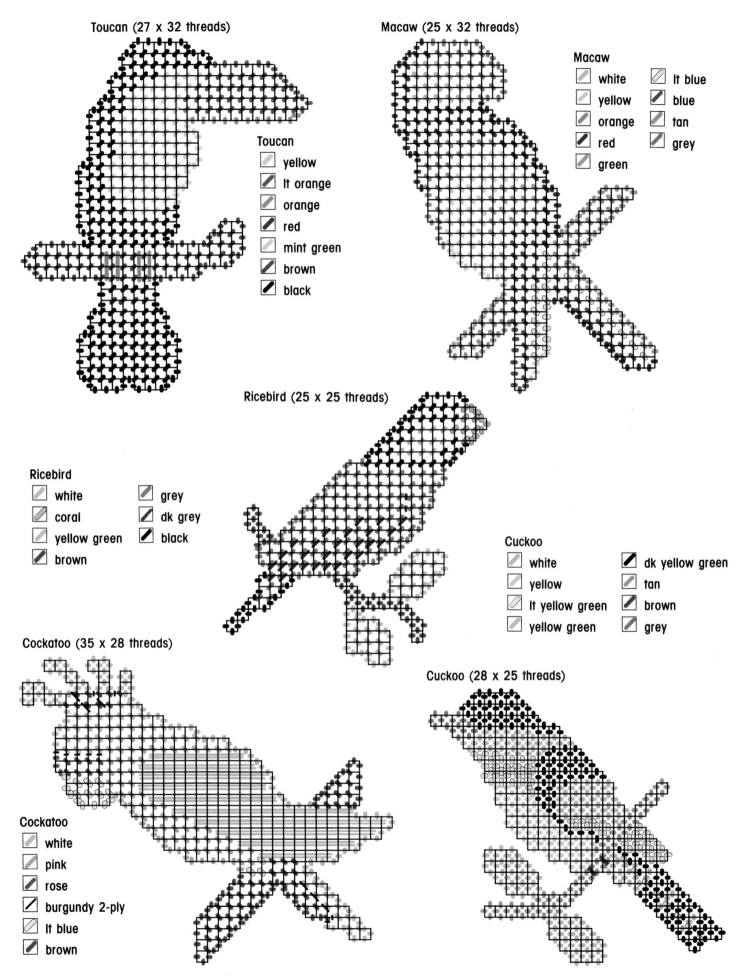

Toucan (27 x 32 threads)

Macaw (25 x 32 threads)

Macaw
white		lt blue	
yellow		blue	
orange		tan	
red		grey	
green			

Toucan
- yellow
- lt orange
- orange
- red
- mint green
- brown
- black

Ricebird (25 x 25 threads)

Ricebird
white		grey
coral		dk grey
yellow green		black
brown		

Cuckoo
white		dk yellow green	
yellow		tan	
lt yellow green		brown	
yellow green		grey	

Cockatoo (35 x 28 threads)

Cuckoo (28 x 25 threads)

Cockatoo
- white
- pink
- rose
- burgundy 2-ply
- lt blue
- brown

WORKING WOMAN'S WEEK

A sense of humor often makes life a little easier when you're juggling the responsibilities of family, career, and housework. This helpful magnet offers a witty suggestion for how to get everything done — hire a maid!

Magnet (35 x 36 threads)

WORKING WOMAN'S WEEK

Size: 5¼"w x 5⅜"h

Supplies: Worsted weight yarn (refer to color key), embroidery floss (refer to color key), one 10½" x 13½" sheet of 7 mesh plastic canvas, #16 tapestry needle, two magnetic strips, and clear-drying craft glue

Stitches Used: Backstitch, Overcast Stitch, and Tent Stitch

Instructions: Follow chart and use required stitches to work Magnet, using six strands of embroidery floss for words. Glue magnetic strips to wrong side of Magnet.

Design by Mary Lou Buttram.

- ✏ red
- ✏ blue
- ✏ brown
- ✏ white embroidery floss
- ✏ black embroidery floss

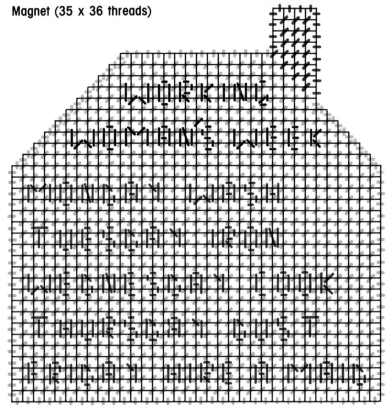

creative crayons

There's no end to the clever projects you can create using these brightly hued crayons! Not only are they attention-getting magnets for your notes and reminders but these colorful pieces also make a cute pencil cup and more.

CREATIVE CRAYONS
Supplies: Worsted weight yarn (refer to color key and photo), one 10½" x 13½" sheet of 7 mesh plastic canvas, one 3" diameter plastic canvas circle, #16 tapestry needle, magnetic strip, and clear-drying craft glue
Stitches Used: Backstitch, Overcast Stitch, and Tent Stitch

MAGNET
Size: 3½"w x ¾"h
Instructions: Follow chart and use required stitches to work Crayon. Glue magnetic strip to wrong side of stitched piece.

FRAME
Size: 3½"w x 3½"h
Instructions: Follow chart and use required stitches to work four Crayons. Refer to photo to glue Crayons together, forming a square. For Back, cut a piece of plastic canvas 23 x 23 threads. For Stand, cut a piece of plastic canvas 10 x 23 threads. Use tan to center and stitch one long edge of Stand to Back. Use tan to tack Back to stitched piece along sides and bottom of Back.

PENCIL CUP
Size: 3½"h x 2½" diameter
Instructions: Follow chart and use required stitches to work eight Crayons. For Cup Side, cut a piece of plastic canvas 45 x 20 threads. Overlap short ends four threads and use tan to join ends, forming a cylinder. For Bottom, cut three threads from 3" diameter plastic canvas circle. Use tan to join Bottom to Side. Refer to photo to glue Crayons to Cup.

Design by Dick Martin.

Crayon (21 x 21 threads)

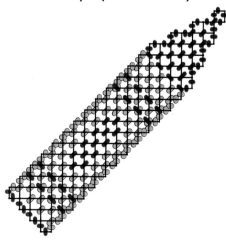

PRETTY PANSIES

Approx Size: 2½"w x 3¾"h each

Supplies: Needloft® Plastic Canvas Yarn or
worsted weight yarn (refer to color key),
one 10½" x 13½" sheet of 7 mesh plastic
canvas, #16 tapestry needle, magnetic
strip, and clear-drying craft glue

Stitches Used: Gobelin Stitch, Overcast
Stitch, and Tent Stitch

Instructions: Follow charts and use
required stitches to work Magnet pieces.
Refer to photo and use yellow to join Pansy
Top to Pansy Bottom. Refer to photo to glue
Pansy to Base. Refer to photo for yarn color
to thread a 10" length of yarn through Base
at ▲'s. Tie yarn in a bow and trim ends.
Glue magnetic strip to wrong side of
Magnet.

Designs by Dick Martin.

NL	COLOR		NL	COLOR
09	rust		45	lilac
20	lemon		57	yellow
24	mint		59	plum
39	eggshell			

Purple Pansy Top
(10 x 10 threads)

Purple Pansy Bottom
(10 x 10 threads)

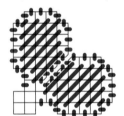

Yellow Pansy Base (16 x 22 threads)

Purple Pansy Base (16 x 22 threads)

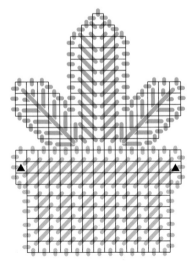

Yellow Pansy Top
(10 x 10 threads)

Yellow Pansy Bottom
(10 x 10 threads)

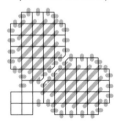

Country Friends

Our perky pig, cow, and goose will charm their way right into your heart.

COUNTRY FRIENDS

Approx Size: 3³/₈"w x 3¹/₂"h each

Supplies: Needloft® Plastic Canvas Yarn or worsted weight yarn (refer to color key), one 10¹/₂" x 13¹/₂" sheet of 7 mesh plastic canvas, #16 tapestry needle, magnetic strip, and clear-drying craft glue

Stitches Used: Backstitch, French Knot, Overcast Stitch, and Tent Stitch

Instructions: Follow chart and use required stitches to work Magnet. Glue magnetic strip to wrong side of Magnet. **For Cow only,** cut 4" of black yarn. Tie yarn in a knot ¹/₂" from one end and separate yarn below knot into plies. Thread loose end of yarn through canvas from front to back at ▲. Secure yarn on wrong side of stitched piece.

NL	COLOR		NL	COLOR
00	black		39	off-white
07	lt pink		56	flesh
12	orange		00	black Fr. Knot
33	dk blue			

Pig (24 x 16 threads)

Cow (23 x 18 threads)

Goose (15 x 21 threads)

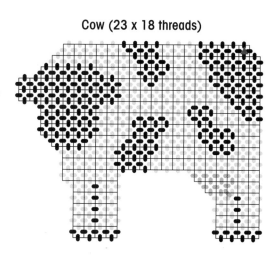

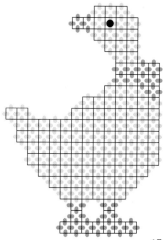

Shipshape Lighthouses

Our shipshape magnets will not only be a beacon of fun, but they'll add nautical appeal to your home, too! The five designs were inspired by traditional seaside lighthouses.

SHIPSHAPE LIGHTHOUSES

Approx Size: 3¼"w x 3¾"h each

Supplies: Worsted weight yarn (refer to color key), green embroidery floss, one 10½" x 13½" sheet of 7 mesh plastic canvas, #16 tapestry needle, magnetic strip, and clear-drying craft glue

Stitches Used: Backstitch, Cross Stitch, Overcast Stitch, and Tent Stitch

Instructions: When color key indicates embroidery floss, use six strands. Follow chart and use required stitches to work Magnet. Glue magnetic strip to wrong side of Magnet.

Designs by Nancy Dorman.

⬜	white
⬜	red
⬜	green
⬜	dk green
⬜	rust
⬜	dk rust
⬜	tan
⬜	brown
⬜	lt grey
⬜	grey
⬜	dk grey
⬜	black
⬜	green embroidery floss

Lighthouse #3 (23 x 25 threads)

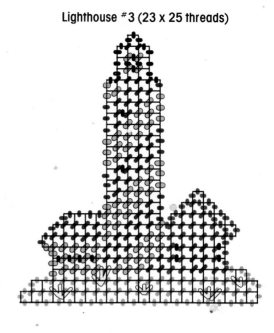

Lighthouse #1 (22 x 27 threads)

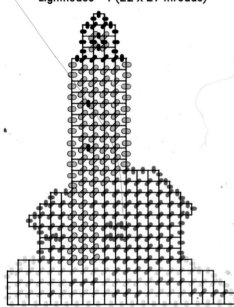

Lighthouse #4 (21 x 25 threads)

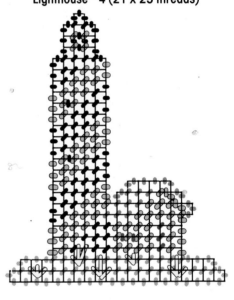

Lighthouse #2 (22 x 24 threads)

Lighthouse #5 (25 x 26 threads)

THANKSGIVING PILGRIMS

Approx Size: 2 1/8"w x 3 1/2"h each

Supplies: Needloft® Plastic Canvas Yarn or worsted weight yarn (refer to color key), six strand embroidery floss (refer to color key), one 10 1/2" x 13 1/2" sheet of 7 mesh plastic canvas, #16 tapestry needle, magnetic strip, and clear-drying craft glue

Stitches Used: Backstitch, French Knot, Overcast Stitch, and Tent Stitch

Instructions: Follow chart and use required stitches to work Magnet. Glue magnetic strip to wrong side of Magnet.

Designs by Dick Martin.

Pilgrim Woman (19 x 19 threads)

Pilgrim Man (21 x 21 threads)

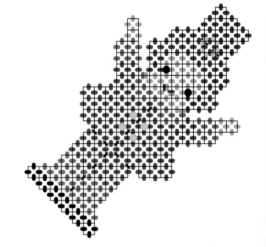

NL	COLOR
✎ 00	black
✎ 07	pink
✎ 13	brown
✎ 38	grey

NL	COLOR
✎ 41	white
✎ 56	flesh
✎ 57	yellow
✎	orange embroidery floss

NL	COLOR
✎	black embroidery floss
● 57	yellow Fr. Knot
●	black embroidery floss Fr. Knot

AUTUMN LEAVES

Echoing the brilliant hues of fall, our leaf-shaped magnets offer a rich harvest of autumn beauty. With three distinct leaves — oak, elm, and maple — these handy items are a lovely way to "fall" into the season.

AUTUMN LEAVES
Approx Size: 2¹/₂"w x 3¹/₂"h each
Supplies: Worsted weight yarn (refer to color key), one 10¹/₂" x 13¹/₂" sheet of 7 mesh plastic canvas, #16 tapestry needle, magnetic strip, and clear-drying craft glue

Stitches Used: Backstitch, Gobelin Stitch, Overcast Stitch, and Tent Stitch
Instructions: Follow chart and use required stitches to work Magnet. Glue magnetic strip to wrong side of stitched piece.

- gold
- orange
- brown
- dk brown

Elm Leaf (19 x 19 threads)

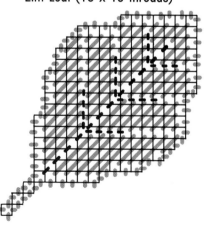

Oak Leaf (19 x 19 threads)

Maple Leaf (18 x 18 threads)

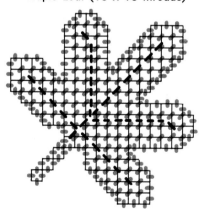

Cheery Cardinal and Chickadee

Cardinals and chickadees are some of nature's most beautiful creatures. These magnets capture the gracefulness of the popular backyard visitors.

CHEERY CARDINAL & CHICKADEE

Approx Size: 3³/₈"w x 3¹/₂"h each

Supplies: Needloft® Plastic Canvas Yarn or worsted weight yarn (refer to color key), one 10¹/₂" x 13¹/₂" sheet of 7 mesh plastic canvas, #16 tapestry needle, magnetic strip, and clear-drying craft glue

Stitches Used: Backstitch, Cross Stitch, Overcast Stitch, and Tent Stitch

Instructions: Follow chart and use required stitches to work Magnet. Complete backgrounds with white Tent Stitches as indicated on chart before working Backstitches. Glue magnetic strip to wrong side of Magnet.

Designs by Joan E. Ray.

NL	COLOR		NL	COLOR
00	black		38	grey
02	red		39	ecru
12	orange		41	white
14	brown		42	dk red
17	gold		56	flesh
29	green			

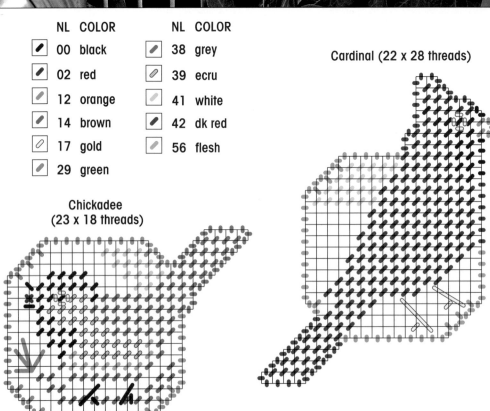

Cardinal (22 x 28 threads)

Chickadee (23 x 18 threads)

Gentle Robin & Bluebird

Hopping and flitting about, the robin and bluebird fascinate us each spring as they prepare to raise their young. Our magnets let you invite these colorful birds into your home.

GENTLE ROBIN & BLUEBIRD

Approx Size: 3⅞"w x 3¾"h each

Supplies: Worsted weight yarn or Needloft® Plastic Canvas Yarn (refer to color key), one 10½" x 13½" sheet of 7 mesh plastic canvas, #16 tapestry needle, magnetic strip, and clear-drying craft glue

Stitches Used: Backstitch, Cross Stitch, Overcast Stitch, and Tent Stitch

Instructions: Follow chart and use required stitches to work Magnet. Complete background with ecru Tent Stitches as indicated on chart, before working Backstitches. Glue magnetic strip to wrong side of Magnet.

Designs by Joan E. Ray.

NL	COLOR	NL	COLOR	NL	COLOR
00	black	16	tan	39	ecru
10	rust	29	dk green	41	white
12	lt rust	32	blue	43	lt brown
14	brown	35	lt blue	53	green
15	dk brown	38	grey	57	gold

Bluebird (28 x 23 threads)

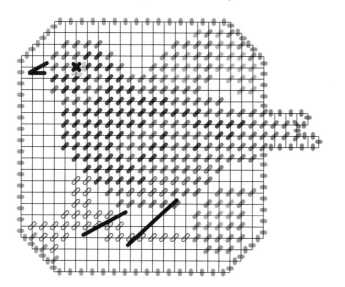

Robin (26 x 26 threads)

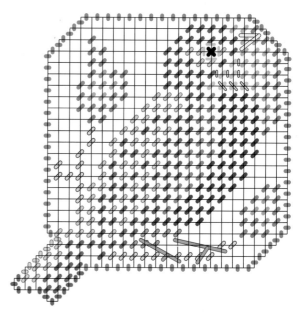

SUMMER FUN

*To remind you of sunny summer days all year long,
we chose this set of five whimsical magnets. There's an adorable frog,
a comical cactus, a brilliant sun, a "Gone Fishin'" sign,
and a garden-fresh ear of corn.*

Fishing
Tournament
Friday

22

SUMMER FUN

Approx Size: 3½"w x 3"h each

Supplies: Worsted weight yarn or Needloft® Plastic Canvas Yarn (refer to color key), one 10½" x 13½" sheet of 7 mesh plastic canvas, #16 tapestry needle, two 6mm moving eyes, magnetic strip, and clear-drying craft glue

Stitches Used: Alicia Lace, Backstitch, Cross Stitch, French Knot, Fringe, Overcast Stitch, and Tent Stitch

Instructions: Follow chart and use required stitches to work Magnet. Glue magnetic strip to wrong side of Magnet. **For Cactus only:** Refer to photo to glue moving eyes to Cactus. **For Corn only:** Trim Fringe to ½" and separate plies. **For Frog only:** Refer to photo to tack Legs to Frog.

NL	COLOR
✎ 00	black
✎ 02	red
✎ 10	brown
✎ 28	green
✎ 29	dk green
✎ 35	blue
✎ 41	white
✎ 57	yellow
◉ 28	green 2-ply Fr. Knot
• 57	yellow Fr. Knot
○ 10	brown Fringe

Sun (19 x 19 threads)

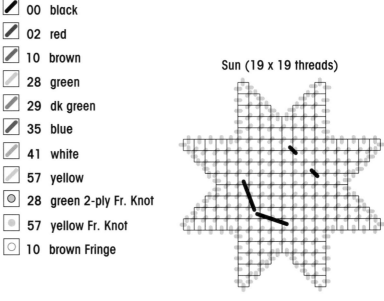

Sun designed by Lisette Rodriguez.

Cactus (20 x 28 threads)

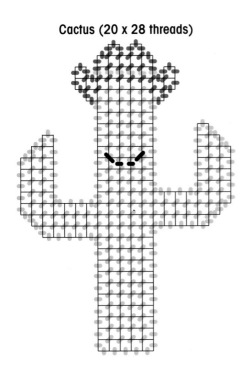

Cactus designed by Juanita J. Simser.

Gone Fishin' (24 x 18 threads)

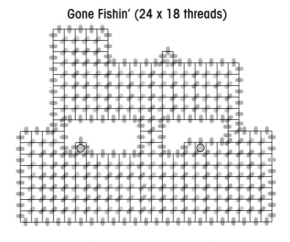

Gone Fishin' designed by Laura Farnham.

Corn (12 x 20 threads)

Frog (42 x 19 threads)

Leg (6 x 8 threads)
(Work 2)

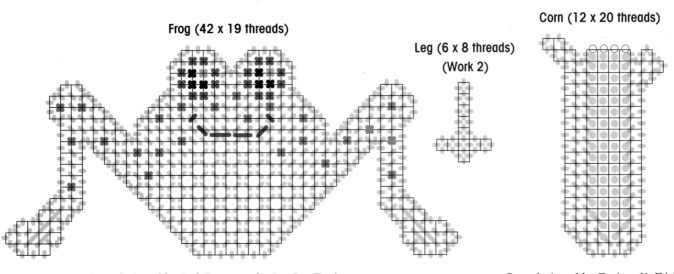

Frog designed by Jack Peatman for LuvLee Designs.

Corn designed by DeAnn K. Dickman.

BEARY SPECIAL MESSAGES

Perfect for posting "honey-do" notes, shopping lists, and other reminders, these sweet magnets will make your messages more "bearable."

BEARY SPECIAL MESSAGES

Size: 2"w x 2⅝"h each

Supplies: Worsted weight yarn (refer to color key), one 10½" x 13½" sheet of 7 mesh plastic canvas, #16 tapestry needle, magnetic strip, 1"h flocked bear, and clear-drying craft glue

Stitches Used: Backstitch, Overcast Stitch, and Tent Stitch

Instructions: Follow chart and use required stitches to work Magnet. (**Note:** Complete background with ecru Tent Stitches as indicated on chart before adding Backstitch.) Refer to photo to glue flocked bear to Magnet. Glue magnetic strip to wrong side of stitched piece.

For Magnet #1 only: Refer to photo to glue Heart to Magnet.

Designs by Joan Bartling.

	ecru		blue
	yellow		green
	red		black 2-ply

Heart (3 x 3 threads)

Magnet #1 (14 x 18 threads)

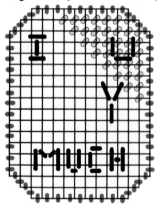

Magnet #2 (14 x 18 threads)

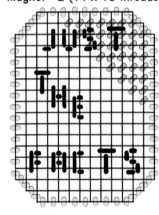

Magnet #3 (14 x 18 threads)

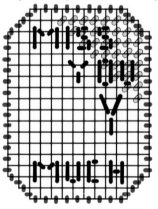

Magnet #4 (14 x 18 threads)

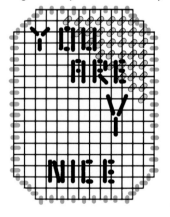

CANISTER ACCENTS

Your favorite cook will enjoy this set of miniature kitchen canister magnets. The attractive, yet functional, accents are ideal for holding grocery lists, meal plans, and even recipes.

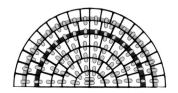

Top/Bottom
(Work 2 of each Size)

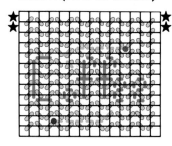

Rim A (15 x 3 threads)

Front A (15 x 13 threads)

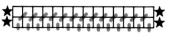

Rim B (18 x 3 threads)

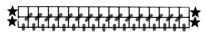

Front B (18 x 16 threads)

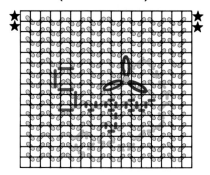

Rim C (21 x 3 threads)

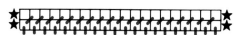

Front C (21 x 19 theads)

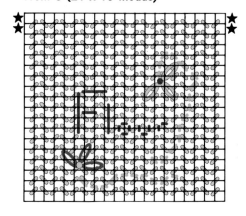

CANISTER ACCENTS

Canister A Size: 1½"w x 2⅜"h
Canister B Size: 2"w x 2½"h
Canister C Size: 2⅜"w x 3¼"h

Supplies: Worsted weight yarn (refer to color key), one 10½" x 13½" sheet of 7 mesh plastic canvas, six 3" diameter plastic canvas circles, #16 tapestry needle, three 10mm pom-poms (refer to photo for colors), magnetic strip, and clear-drying craft glue

Stitches Used: Backstitch, French Knot, Lazy Daisy Stitch, Overcast Stitch, and Tent Stitch

Instructions: Follow charts and use required stitches to work Canister Magnet pieces. For Back A, cut a piece of plastic canvas 9 x 13 threads. For Back B, cut a piece of plastic canvas 12 x 16 threads. For Back C, cut a piece of plastic canvas 15 x 19 threads. (**Note:** Backs are not worked.) Match ★'s to place Rim on Front. Use white to join Rim and Front to Back along long edges of Back. Use yarn color to match Rim to join Top to Rim, Front, and Back. Use white to join Bottom to Front and Back. Refer to photo to glue pom-poms to Tops. Glue magnetic strip to Back.

Designs by Jennifer Schmidt.

▱ white	⊙ yellow Fr. Knot	▱ blue Lazy Daisy	
▱ red	● red Fr. Knot	▱ cutting line for Top/Bottom A	
▱ blue	● blue Fr. Knot	▱ cutting line for Top/Bottom B	
▱ green 2-ply	▱ yellow Lazy Daisy		
▱ dk green	▱ red Lazy Daisy		

FROSTY FRIENDS

Embellished with glitter and tiny beads, our Mr. and Mrs. Snowman magnets are simply adorable. What a delightful way to display your favorite Yuletide recipes, Christmas wish lists, or holiday greeting cards!

FROSTY FRIENDS

Size: 3¹⁄₂"w x 4"h each

Supplies: Worsted weight yarn (refer to color key), one 10¹⁄₂" x 13¹⁄₂" sheet of 7 mesh clear plastic canvas, one 10¹⁄₂" x 13¹⁄₂" sheet of 7 mesh brown plastic canvas, #16 tapestry needle, Mill Hill® black seed beads, clear glitter, ¹⁄₂" x 8" piece of fabric, sewing thread, beading needle, magnetic strip, and clear-drying craft glue

Stitches Used: Gobelin Stitch, Overcast Stitch, and Tent Stitch

Instructions: Cut Body and Hat from clear plastic canvas. Cut Arms from brown plastic canvas. Follow charts and use required stitches to work magnet pieces. Refer to photo to sew beads to Body. Spread a thin layer of glue on Body; sprinkle with glitter. Refer to photo to glue Arms to Body. Glue magnetic strip to wrong side of stitched piece. **For Mr. Snowman only,** refer to photo to glue Hat to Body. **For Mrs. Snowman only,** refer to photo to tie fabric around neck and trim ends.

Designs by Kelli DePasse Smith.

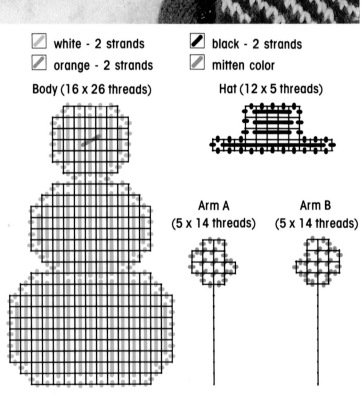

	white - 2 strands		black - 2 strands
	orange - 2 strands		mitten color

Body (16 x 26 threads)

Hat (12 x 5 threads)

Arm A (5 x 14 threads)

Arm B (5 x 14 threads)

Fetching Trio

This fetching trio is sure to make you smile every time you post your shopping list. The magnetic pups and hydrant also make doggone good gifts.

FETCHING TRIO

Scottie Size: 3"w x 2½"h

Hydrant Size: 2"w x 3"h

Supplies: Worsted weight yarn (refer to color key and photo), black embroidery floss, one 10½" x 13½" sheet of 7 mesh plastic canvas, #16 tapestry needle, two 4mm rhinestones, 18" of ⅝"w striped ribbon (refer to photo), magnetic strip, and clear-drying craft glue

Stitches Used: Backstitch, Gobelin Stitch, Overcast Stitch, Tent Stitch, and Turkey Loop

Instructions: Use six strands of embroidery floss for Backstitches. Follow chart and use required stitches to work Magnet. Glue magnetic strip to wrong side of Magnet.

For Scottie Magnet only: Tie a 9" length of ribbon in a bow and trim ends. Refer to photo to glue bow to Magnet. Refer to photo to glue rhinestone to Magnet. Clip Turkey Loops and separate into plies.

Scottie Magnet design by Terre Uram.

✒ Scottie color - 3 yds	✒ black embroidery floss - 1 yd
✒ red - 3 yds	⊙ Scottie color Turkey Loop

Hydrant Magnet (13 x 20 threads)

Scottie Magnet (21 x 17 threads)

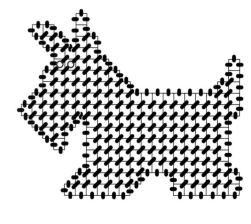

Flowers for You

You can create your own bouquet in no time with this collection of magnets. Whether you stitch one or all of them, these blooms will make great gifts — for friends or for yourself!

FLOWERS FOR YOU

Size: 2³/₄"w x 3¹/₄"h each
Supplies: Worsted weight yarn or Needloft® Plastic Canvas Yarn (refer to color key), six strand blue embroidery floss, one 10¹/₂" x 13¹/₂" sheet of 7 mesh plastic canvas, #16 tapestry needle, magnetic strip, and clear-drying craft glue
Stitches Used: Backstitch, French Knot, Overcast Stitch, and Tent Stitch
Instructions: Follow chart and use required stitches to work Magnet. Glue magnetic strip to wrong side of Magnet.

NL	COLOR		NL	COLOR		NL	COLOR
03	vy dk rose		20	lt yellow		53	green
05	rose		26	lt green		59	purple
06	dk rose		27	dk green			blue embroidery floss
07	lt rose		33	blue		05	rose Fr. Knot
12	peach		34	lt blue		06	dk rose Fr. Knot
16	beige		44	lt purple		17	dk yellow Fr. Knot
19	yellow		47	lt peach		46	dk purple Fr. Knot

Flower #1 (14 x 23 threads)

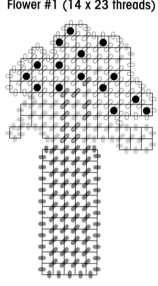

Flower #2 (24 x 16 threads)

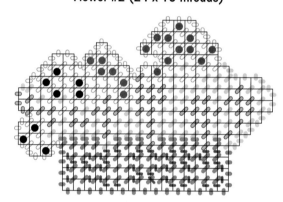

Flower #3 (21 x 22 threads)

Flower #4 (17 x 22 threads)

Flower #5 (17 x 28 threads)

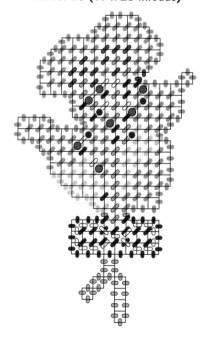

29

KITTEN CABOODLE

With three playful poses, these kitty magnets are purr-fect for holding your notes and reminders!

KITTEN CABOODLE

Cat #1 Size: $2^1/_4$"w x $4^1/_4$"h
Cat #2 Size: $2^7/_8$"w x $3^3/_8$"h
Cat #3 Size: $4^1/_2$"w x $2^1/_8$"h

Supplies: Worsted weight yarn (refer to color key), one $10^1/_2$" x $13^1/_2$" sheet of 7 mesh plastic canvas, #16 tapestry needle, one 4mm gold jingle bell, three 2" lengths of black cloth-covered floral wire, two 8" lengths of $^1/_{16}$"w red satin ribbon, magnetic strip, and clear-drying craft glue

Stitches Used: Overcast Stitch and Tent Stitch

Instructions: Follow chart and use required stitches to work magnet. Glue magnetic strip to wrong side of Magnet.

For Cat #1 only: Thread ribbon through jingle bell. Refer to photo to glue ends of ribbon to wrong side of Magnet.

For Cat #2 only: Tie one length of ribbon in a bow; trim ends. Refer to photo to glue bow to Magnet.

For Cat #3 only: Thread black cloth-covered floral wire through Cat at ✳'s. Secure wire with a dot of glue on wrong side of Magnet.

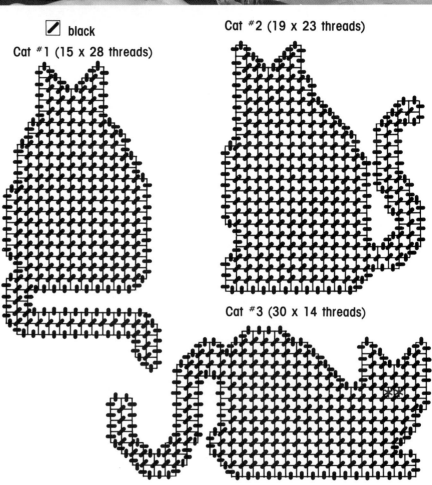

■ black

Cat #1 (15 x 28 threads)

Cat #2 (19 x 23 threads)

Cat #3 (30 x 14 threads)

POODLE PARADE

Accented with rhinestones and ribbons, these cute magnets bring back fun memories of the 1950's when poodle skirts, soda fountains, and sock hops were really hip!

POODLE PARADE

Size: 3¼"w x 2"h

Supplies: Worsted weight yarn (refer to photo), one 10½" x 13½" sheet of 7 mesh plastic canvas, #16 tapestry needle, two 7mm moving eyes, one ¼" black pom-pom, two 4½mm crystal rhinestones, two 10" lengths of ⅛"w satin ribbon, pink felt, magnetic strip, and clear-drying craft glue

Stitches Used: Backstitch, Overcast Stitch, and Turkey Loop

Instructions: Follow chart and use required stitches to work Poodle. For tongue, cut a small oval from pink felt and glue to wrong side of Poodle. Refer to photo to glue moving eyes and pom-pom to Poodle. Tie each length of ribbon in a bow and trim ends. Glue one rhinestone to each bow. Glue bows to Poodle. Glue magnetic strip to wrong side of Magnet.

Design by Elizabeth L. Adams.

Poodle (18 x 18 threads)

▨ poodle color   ⊙ poodle color Turkey Loop

TULIP TIME

You can't tiptoe through these tulips, but you'll love planting the colorful reminders of spring on your refrigerator door — or anywhere you want! Use them to hold important notes, favorite snapshots, or your child's latest artwork.

TULIP TIME

Size: 2"w x 4"h x 5/8"d

Supplies: Worsted weight yarn (refer to color key), one 10½" x 13½" sheet of 7 mesh clear plastic canvas, #16 tapestry needle, magnetic strip, and clear-drying craft glue

Stitches Used: Backstitch, Overcast Stitch, and Tent Stitch

Instructions: Follow charts and use required stitches to work Tulip magnet pieces, leaving stitches in shaded area unworked. (**Note:** Back is not worked.) Use yarn color to match stitching area for all joining. Match ◆'s and work stitches in shaded areas to join Petal to Front. Match ✖'s to join Flowerpot Edge to Front between ✖'s. Tack Leaf to Front at ▲'s. Join Back to Front along unworked edges. Glue magnetic strip to Back.

Design by Dick Martin.

	green		brown
	dk green		Tulip color

Front/Back
(24 x 24 threads) (Cut 2, Work 1)

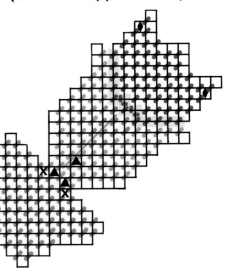

Flowerpot Edge
(11 x 11 threads)

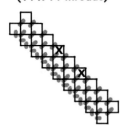

Leaf (8 x 8 threads)

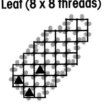

Petal (11 x 11 threads)

Noteworthy Magnets

Express your sentiments loud and clear with these noteworthy signs! Whether you're highlighting an urgent message or sending "no smoking" signals, these magnets will get the job done.

NOTEWORTHY MAGNETS
"See This" Size: 5³/₈"w x 1⁷/₈"h
No Smoking Size: 3¹/₂"w x 3¹/₂"h
Supplies: Worsted weight yarn or Needloft® Plastic Canvas Yarn (refer to color key), embroidery floss (refer to color key), one 10¹/₂" x 13¹/₂" sheet of 7 mesh plastic canvas, #16 tapestry needle, magnetic strip, and clear-drying craft glue
Stitches Used: Backstitch, Overcast Stitch, and Tent Stitch
Instructions: Use six strands of embroidery floss for Backstitch. Follow chart and use required stitches to work Magnet. Glue magnetic strip to wrong side of stitched piece.

"See This" design by Mary Fitkowski.
No Smoking design by Esther L. Metras.

NL	COLOR		NL	COLOR
02	red		57	yellow
13	tan			white embroidery floss
38	grey			black embroidery floss
41	white			

"See This" (37 x 13 threads)

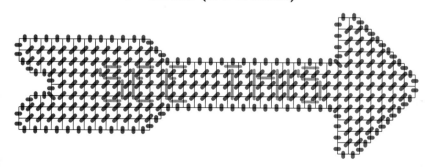

No Smoking (24 x 24 threads)

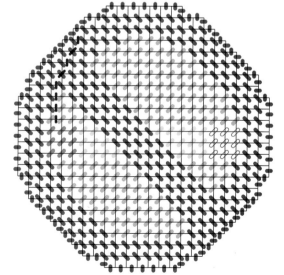

HOLIDAY LIGHTS

Light up the holiday season with these symbolic magnets — a menorah for Chanukah and an Advent wreath for Christmas. The candles on these designs are "lighted" by placing a flame magnet over each taper at the appropriate time during the celebration. Happy Chanukah and Merry Christmas to each of you!

HOLIDAY LIGHTS
Menorah Size: 5½"w x 5½"h
Advent Wreath Size: 4"w x 5"h
Supplies: Worsted weight yarn (refer to color key), one 10½" x 13½" sheet of 7 mesh plastic canvas, #16 tapestry needle, Kreinik ⅛"w metallic gold ribbon, magnetic strip, and clear-drying craft glue

Stitches Used: French Knot, Overcast Stitch, and Tent Stitch
Instructions: Follow charts and use required stitches to work Menorah and nine Flames or Advent Wreath and four Flames. Glue magnetic strips to wrong sides of stitched pieces.

Designs by Mary K. Perry.

⊘ white		⊘ green	
⊘ yellow		⊘ dk green	
⊘ orange		⊘ metallic gold ribbon	
⊘ pink		⊙ red 2-ply Fr. Knot	
⊘ purple			

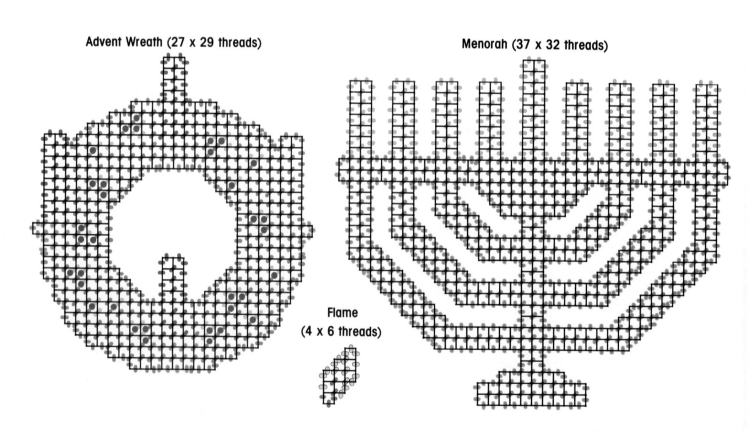

Advent Wreath (27 x 29 threads)

Menorah (37 x 32 threads)

Flame
(4 x 6 threads)

BOOGIE-WOOGIE FRUIT

Our high-stepping fruit magnets will dance their way right into your heart!
With open arms and cheerful smiles, they'll make happy additions to your home.

BOOGIE-WOOGIE FRUIT

Approx Size: 4¹/₂"w x 4¹/₂"h each

Supplies: Worsted weight yarn (refer to color key), black six strand embroidery floss, one 10¹/₂" x 13¹/₂" sheet of 7 mesh plastic canvas, #16 tapestry needle, ¹/₂" pom-pom (refer to photo for color), 15mm oval moving eyes, magnetic strip, and clear-drying craft glue

Stitches Used: Backstitch, French Knot, Gobelin Stitch, Overcast Stitch, and Tent Stitch

Instructions: Follow chart and use required stitches to work Magnet. Refer to photo to glue moving eyes and pom-pom to Magnet. Glue magnetic strip to wrong side of Magnet.

Designs by Sheri Lautenschlager.

	white		green
	yellow		tan
	orange		brown
	pink		black
	red		black embroidery floss
	lt green	•	white Fr. Knot

Watermelon (36 x 24 threads)

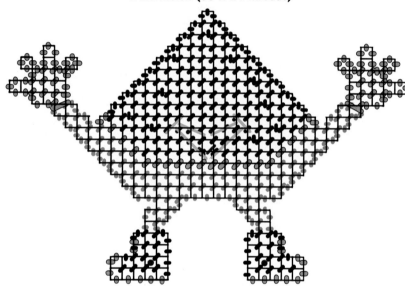

Banana (25 x 40 threads)

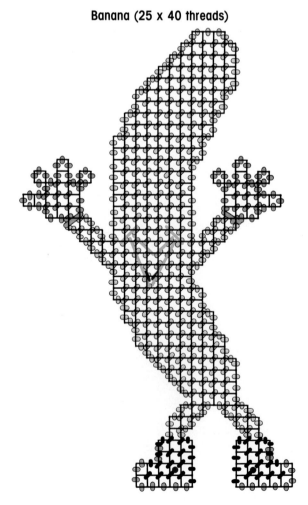

Orange (33 x 25 threads)

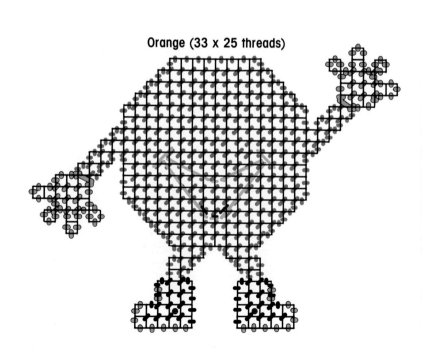

Apple (33 x 28 threads)

Pear (28 x 30 threads)

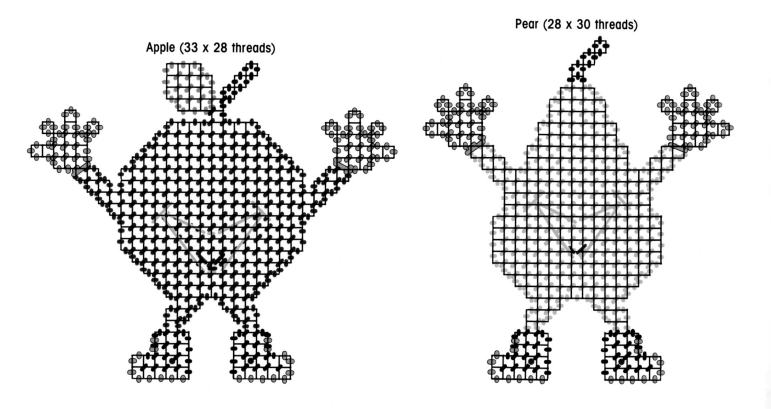

Pineapple (32 x 31 threads)

Strawberry (30 x 27 threads)

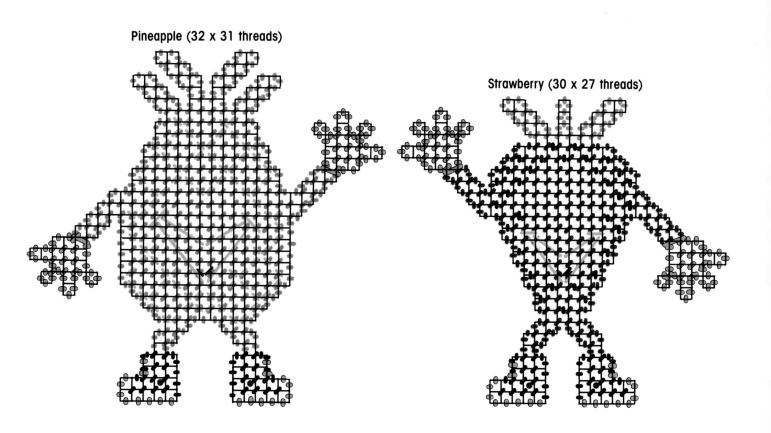

37

DIET LIGHT

Here's a bright idea for letting others know you have your diet under control. This cute on/off magnet will tell them ''Don't tempt me!'' or ''Don't scold me — I'm eating this on purpose!''

DIET LIGHT
Size: 2⁷/₈"w x 5¹/₂"h
Supplies: Worsted weight yarn (refer to color key), one 10¹/₂" x 13¹/₂" sheet of 7 mesh plastic canvas, #16 tapestry needle, 6" of ¹/₁₆"w yellow satin ribbon, magnetic strip, and clear-drying craft glue
Stitches Used: Backstitch, Overcast Stitch, and Tent Stitch
Instructions: Follow charts and use required stitches to work Magnet pieces. Tie ribbon in a knot close to one end. Refer to photo to thread loose end of ribbon under stitches on wrong side of Light Bulb. Knot loose end of ribbon ¹/₂" below Light Bulb and trim ribbon close to knot. Place knotted ribbon between wrong sides of On and Off. Use grey to join On to Off. Glue magnetic strip to wrong side of Light Bulb.

Design by Sharon Adkins.

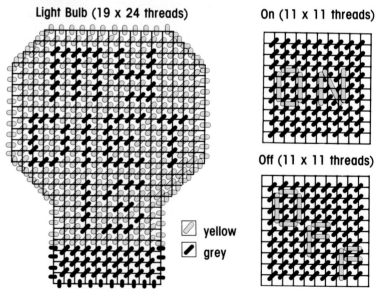

Light Bulb (19 x 24 threads)

On (11 x 11 threads)

Off (11 x 11 threads)

| | yellow |
| | grey |

Tropical Fish

These "sea-sonal" accents offer a fun way to decorate your bathroom, sun-room, or any room you want to give a tropical feeling. Our three exotic fish are certain to add a splash of color no matter where you display them!

TROPICAL FISH

Approx Size: 3"w x 3"h each
Supplies: Needloft® Plastic Canvas Yarn or worsted weight yarn (refer to color key), one 10½" x 13½" sheet of 7 mesh plastic canvas, #16 tapestry needle, 7 mm moveable eyes, magnetic strip, and clear-drying craft glue
Stitches Used: Backstitch, Overcast Stitch, and Tent Stitch
Instructions: Follow chart and use required stitches to work Magnet. Refer to photo to glue moving eye to Fish. Glue magnetic strip to wrong side of Magnet.

Designs by Vivian Mitchell.

Fish B (25 x 19 threads)

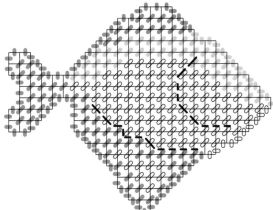

NL	COLOR
✎ 00	black - 1 yd
⬚ 28	Christmas green - 4 yds
✎ 32	royal - 3 yds
✎ 46	purple - 2 yds
⬚ 57	yellow - 4 yds
⬚ 58	bright orange - 1 yd

Fish A (24 x 22 threads)

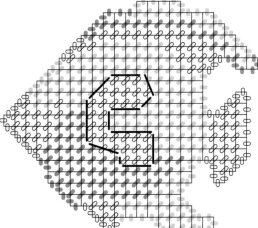

Fish C (23 x 15 threads)

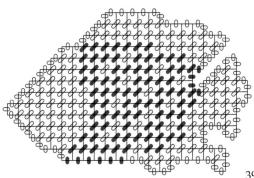

Football Helmet (13 x 13 threads)

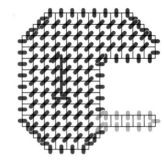

Football (17 x 11 threads)

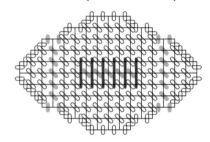

Basketball (14 x 14 threads)

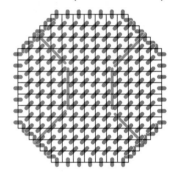

Sporty Magnets

You'll score big points with these sporty magnets! Whether for your favorite athlete or yourself, these quick-to-stitch designs are real winners.

SPORTY MAGNETS

Approx Size: 1⁷/₈"w x 2¹/₂"h each
Supplies: Worsted weight yarn (refer to color key), embroidery floss (refer to color key), one 10¹/₂" x 13¹/₂" sheet of 7 mesh plastic canvas, #16 tapestry needle, magnetic strip, and clear-drying craft glue
Stitches Used: Backstitch, Overcast Stitch, and Tent Stitch

Instructions: Use six strands of embroidery floss for Backstitch. Follow chart and use required stitches to work Magnet pieces. Glue magnetic strip to wrong side of Magnet. Refer to photo to glue pieces together.

Designs by Karen Hanley.

Soccer Ball (13 x 13 threads)

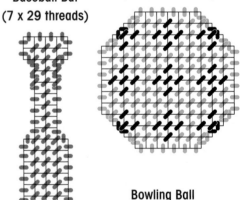

Baseball Bat (7 x 29 threads)

Bowling Ball (11 x 11 threads)

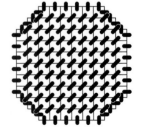

NL	COLOR
00	black
02	red
10	orange
15	brown
19	yellow
38	grey
41	white
43	tan
	white embroidery floss
	black embroidery floss

Tennis Ball (3 x 3 threads)

Baseball (5 x 5 threads)

Tennis Racket (10 x 22 threads)

Bowling Pin (8 x 23 threads)

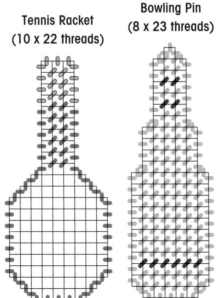

Puppy Time

Dog lovers will get a kick out of our cute pups. The three different poses — playful, sleepy, and inquisitive — are sure to warm your heart.

PUPPY TIME

Approx Size: 4½"w x 3"h each

Supplies: Needloft® Plastic Canvas Yarn or worsted weight yarn (refer to color key), one 10½" x 13½" sheet of 7 mesh plastic canvas, #16 tapestry needle, magnetic strip, and clear-drying craft glue

Stitches Used: Backstitch, Cross Stitch, Gobelin Stitch, Overcast Stitch, and Tent Stitch

Instructions: Follow chart and use required stitches to work Magnet. Glue magnetic strip to wrong side of Magnet.

Designs by Carole L. Rodgers.

NL	COLOR		NL	COLOR
00	black - 2 yds		41	white - 2 yds
01	red - 1 yd		43	camel - 9 yds
14	cinnamon - 4 yds			

Puppy B (36 x 15 threads)

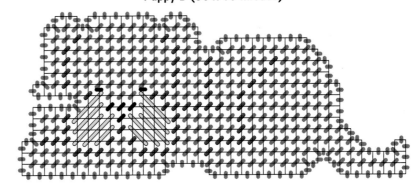

Puppy C (35 x 18 threads)

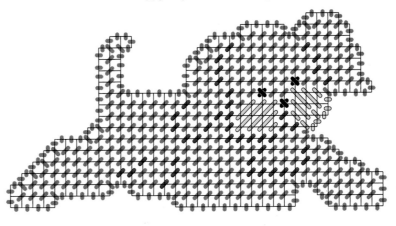

Puppy A (20 x 25 threads)

Country Images

This charming set features lots of country images. Heartwarming and wholesome, the delightful collection includes a teddy bear, a milk bottle, a cat, and more. There's even a pineapple, the traditional symbol of hospitality. Whether you stitch one or all of them, these magnets are sure to become favorites.

COUNTRY IMAGES

Approx Size: 2³/₄"w x 3³/₈"h each

Supplies: Worsted weight yarn or Needloft® Plastic Canvas Yarn (refer to color key), one 10¹/₂" x 13¹/₂" sheet of 7 mesh plastic canvas, #16 tapestry needle, magnetic strip, and clear-drying craft glue

Stitches Used: Backstitch, Cross Stitch, Gobelin Stitch, Mosaic Stitch, Overcast Stitch, Scotch Stitch Variation, and Tent Stitch

Instructions: Follow chart and use required stitches to work Magnet. Glue magnetic strip to wrong side of Magnet.

Designs by Teal Lee Elliott.

NL	COLOR
03	dk rose
05	rose
17	gold
29	green
31	dk blue
33	blue
39	ecru
42	red
	dk rose*
	red*

*Use 2-ply yarn.

Milk Bottle (15 x 26 threads)

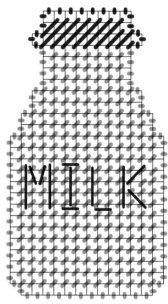

Cat (18 x 24 threads)

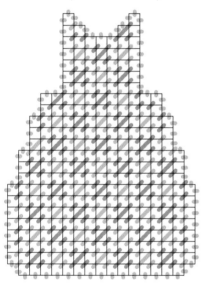

Bear (18 x 25 threads)

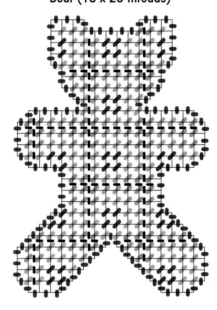

Pineapple (14 x 23 threads)

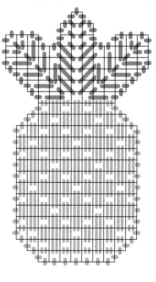

Quilt Square (20 x 20 threads)

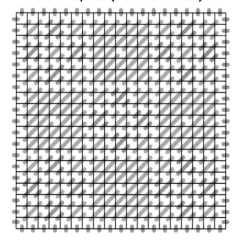

Star (20 x 20 threads)

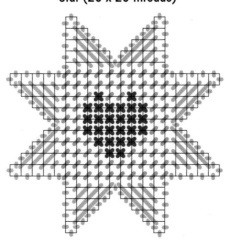

House (21 x 22 threads)

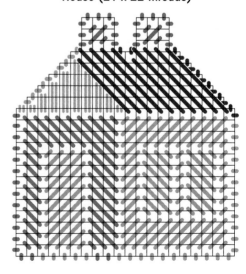

Little Piggies

You'll go hog wild over these little piggies! Sporting colorful bow ties, these whimsical accents are a fun way to brighten a kitchen. What a novel gift for someone who likes to ham it up!

LITTLE PIGGIES

Size: 3¹/₈"w x 3¹/₄"h

Supplies: Needloft® Plastic Canvas Yarn or worsted weight yarn (refer to color key), one 10¹/₂" x 13¹/₂" sheet of 7 mesh plastic canvas, #16 tapestry needle, magnetic strip, and clear-drying craft glue

Stitches Used: Backstitch, Cross Stitch, Gobelin Stitch, Overcast Stitch, and Tent Stitch

Instructions: Follow charts and use required stitches to work Magnet pieces. Use peach to join short sides of Nose Side, forming a circle. Use peach to join Nose Front to Nose Side along unworked edges. Use peach and match ■'s to tack Nose to Front. Use peach and match ✳'s to join Ear A to Front. Use peach and match ▲'s to join Ear B to Front. Match ★'s to glue Streamers to Front. For each Bow Loop, match wrong sides and ✕'s. Use bow color to join ends of Bow Loop together. Refer to photo and match ◆'s to glue Bow Loops to Streamers. Glue magnetic strip to wrong side of Magnet.

Design by Jack Peatman for LuvLee Designs.

NL	COLOR
00	black
07	pink
47	peach
	bow color

Nose Front (6 x 6 threads)

Ear A (6 x 7 threads)

Ear B (6 x 7 threads)

Bow Loop (19 x 4 threads) (Work 2)

Streamers (12 x 7 threads)

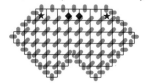

Front (18 x 19 threads)

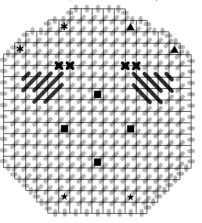

Nose Side (16 x 3 threads)

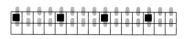

44

SPRING BONNETS

These dainty bonnets recall the days when ladies gathered colorful blossoms to add a touch of spring to their ensembles. Fashioned with satin raffia and adorned with ribbons and flowers, the pretty magnets hold a secret cache of potpourri.

SPRING BONNETS

Size: 1/2"h x 3" diameter

Supplies: Darice® tan Straw Satin, one 10 1/2" x 13 1/2" sheet of 7 mesh plastic canvas, two 3" diameter plastic canvas circles, #16 tapestry needle, 18" of 1/4"w satin ribbon, potpourri, small dried flowers, magnetic strip, and clear-drying craft glue

Stitches Used: Backstitch, Cross Stitch, Gobelin Stitch, and Overcast Stitch

Instructions: For Top, trim four threads from one plastic canvas circle. Follow charts and use required stitches to work magnet pieces, leaving stitches in shaded area unworked. Match **X**'s and work stitches in shaded area to join ends of Side, forming a circle. Join Side to Brim along placement line. Place potpourri inside and join Top to Sides. Refer to photo to decorate hats with ribbon and flowers. Glue magnetic strip to wrong side of stitched piece.

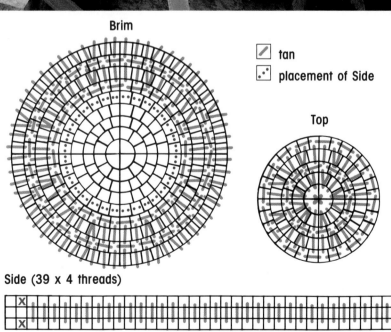

Brim

☑ tan

⊡ placement of Side

Top

Side (39 x 4 threads)

Yuletide Favorites

These Yuletide favorites embody the things we like best about Christmas. The angel heralds the good news of Jesus' birth, the dove symbolizes peace on earth, and Santa represents kindhearted generosity.

YULETIDE FAVORITES

Approx Size: 2³/₄"w x 3"h each

Supplies: Needloft® Plastic Canvas Yarn or worsted weight yarn (refer to color key), embroidery floss (refer to color key), metallic gold braid, one 10¹/₂" x 13¹/₂" sheet of 7 mesh plastic canvas, #16 tapestry needle, magnetic strip, and clear-drying craft glue

Stitches Used: Backstitch, Cross Stitch, French Knot, Gobelin Stitch, Overcast Stitch, and Tent Stitch

Instructions: Use six strands of embroidery floss. Follow chart and use required stitches to work Magnet. Glue magnetic strip to wrong side of Magnet.

Designs by Dick Martin.

NL	COLOR
02	Christmas red
08	baby pink
10	sundown
27	holly
41	white
56	flesh tone

NL	COLOR
	metallic gold braid
	metallic gold braid*
	orange embroidery floss
02	red Fr. Knot
	metallic gold braid Fr. Knot
	black embroidery floss Fr. Knot

*Use 2 strands of braid.

Dove (18 x 18 threads)

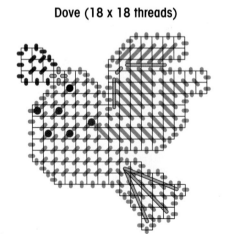

Angel (17 x 17 threads)

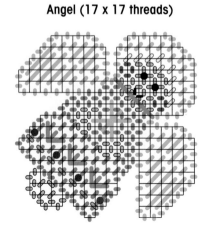

Santa (17 x 18 threads)

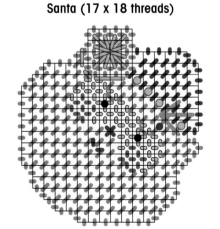

Christmas Critters

Have yourself a merry little Christmas with these magnetic cuties! Holding a youngster's artwork or important messages, the tiny trio is a fun way to spread the holiday spirit. They also make adorable gift tags, tree trims, and wreath accents.

CHRISTMAS CRITTERS

Approx Size: 2¼"w x 3½"h each

Supplies: Worsted weight yarn or Needloft® Plastic Canvas Yarn (refer to color key), one 10½" x 13½" sheet of 7 mesh plastic canvas, #16 tapestry needle, magnetic strip, clear-drying craft glue, and nylon line (for Mouse only)

Stitches Used: Backstitch, Cross Stitch, French Knot, Gobelin Stitch, Overcast Stitch, Tent Stitch, and Turkey Loop

Instructions: Follow chart and use required stitches to work Magnet. Glue magnetic strip to wrong side of Magnet. **For Mouse only:** Thread three 3" lengths of nylon line through canvas at ♦'s. Secure nylon line with a dot of glue on wrong side of Magnet.

✏ white	✏ It grey		
✏ orange	✏ grey		
✏ pink	✏ black		
✏ red	● black Fr. Knot		
✏ green	○ red Turkey Loop		
✏ It brown	○ green Turkey Loop		
✏ brown			

Mouse (16 x 24 threads)

Penguin (16 x 24 threads)

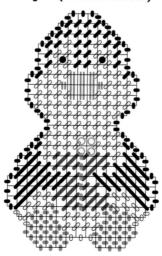

Teddy Bear (14 x 24 threads)

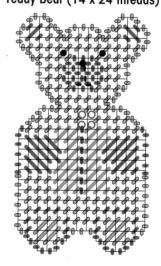

47

SUMMERTIME SNACKS

Now you can indulge in your favorite summertime snacks without gaining an ounce — just ''cook up'' a batch of these junk food magnets! You'll want to stitch extras to share with a friend who's dieting, too.

SUMMERTIME SNACKS

Approx Size: 3⅞"w x 2⅝"h each

Supplies For Entire Set: Worsted weight yarn (refer to color keys), two 10½" x 13½" sheets of 7 mesh plastic canvas, two 3" diameter plastic canvas circles, one 4" diameter plastic canvas circle, #16 tapestry needle, metallic silver ribbon, aluminum foil, magnetic strips, and clear-drying craft glue

Stitches Used For Entire Set: Backstitch, French Knot, Gobelin Stitch, Overcast Stitch, Tent Stitch, and Turkey Loop

Instructions: Follow charts and use required stitches to work magnet pieces, leaving stitches in shaded areas unworked. Glue magnetic strip to wrong side of assembled stitched piece.

For Gum Only: Use yellow for all joining. Join Gum Wrapper Front and Back to Gum Wrapper Sides along long edges. Join Bottom to Front, Back, and Sides. Refer to photo to wrap foil around ends of Gum. Insert Gum into Gum Wrapper.

For Taco Only: Use one 4" diameter plastic canvas circle for Taco Shell. Use half of a 3" diameter plastic canvas circle for Taco Middle. Turn Taco Middle over to work Turkey Loops in shaded areas. Refer to photo to insert Taco Middle into Taco Shell. Use tan to tack Taco Shell closed.

For Pizza Only: Use one 3" diameter plastic canvas circle. Refer to photo to tack 1" of 1-ply ecru yarn to wrong side of Pizza and Pizza Slice. Refer to photo to fray yarn.

Gum design by Betty Flynn.
Hot Dog design by Sue Baker.
Pizza design by Ann Schoch.
Popsicle™ design by Elfie Moore.
Sundae design by Julie Deren.
Taco design by Catherine Bihlmaier.

PIZZA

- ▨ ecru
- ▧ tan
- ▨ red
- • brown Fr. Knot

POPSICLE

- ▨ popsicle color
- ▧ tan

Popsicle (10 x 28 threads)

GUM

- ▨ yellow
- ▧ red 2-ply

Gum Wrapper Front/Back
(20 x 6 threads) (Cut 2, Work 1)

Gum Wrapper Side
(20 x 3 threads) (Work 2)

Gum Wrapper Bottom (6 x 3 threads)

Gum (14 x 5 threads) (Work 2)

Pizza

Pizza Slice

TACO

- ▧ red
- ▧ tan
- ⊙ yellow Turkey Loop*
- ⊙ green Turkey Loop
- ⊙ brown Turkey Loop*
- * Denotes 1" long loops.

Taco Middle

Taco Shell

HOT DOG

- ▨ yellow
- ▧ red
- ▧ dk tan

Hot Dog (22 x 10 threads)

SUNDAE

- ▨ white
- ▧ brown
- ▧ red
- ▧ metallic silver ribbon

Sundae (22 x 23 threads)

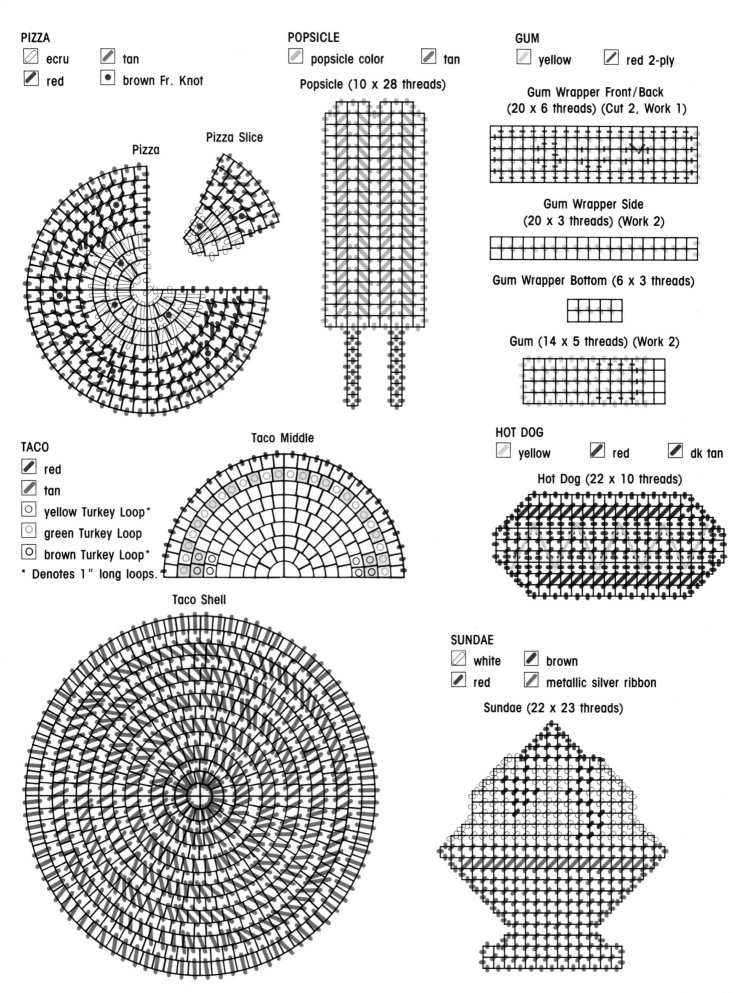

49

Whimsical Trio

Our friendly frog, ladybug, and bumblebee will make it fun to post important reminders close at hand.

WHIMSICAL TRIO

Approx Size: 2³/₈"w x 2⁵/₈"h each

Supplies: Needloft® Plastic Canvas Yarn or worsted weight yarn (refer to color key), one 10¹/₂" x 13¹/₂" sheet of 7 mesh plastic canvas, #16 tapestry needle, magnetic strip, and clear-drying craft glue

Stitches Used: Backstitch, Cross Stitch, French Knot, Gobelin Stitch, Overcast Stitch, and Tent Stitch

Instructions: Follow chart and use required stitches to work Magnet. Glue magnetic strip to wrong side of Magnet. **For Bee and Ladybug only,** wrap black yarn tightly around antennae to cover canvas.

Bee (16 x 20 threads)

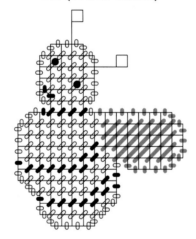

Frog (18 x 18 threads)

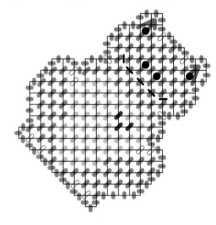

Ladybug (15 x 15 threads)

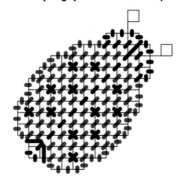

NL	COLOR		NL	COLOR
✎ 00	black		⬚ 29	dk green
✎ 02	red		✎ 41	white
✎ 22	lt green		⬚ 57	yellow
✎ 23	green		● 00	black Fr. Knot

Leave a Note

Handy additions to your message center, our mailbox and flag magnets will help your family keep in touch.

LEAVE A NOTE
Mailbox Size: 6¼"w x 3½"h
Flag Size: 1¾"w x 3⅝"h
Supplies: Needloft® Plastic Canvas Yarn or worsted weight yarn (refer to color key), black embroidery floss, one 10½" x 13½" sheet of 7 mesh plastic canvas, #16 tapestry needle, ⅝"l paper fastener, magnetic strip, and clear-drying craft glue
Stitches Used: Backstitch, Gobelin Stitch, Mosaic Stitch, Overcast Stitch, and Tent Stitch
Instructions: Use six strands of embroidery floss for Backstitch. Follow charts and use required stitches to work Mailbox pieces. Use blue for all joining. Match ★'s and ◆'s to join Side to Bottom. Match ✳'s to join Back to Bottom and Side. To attach Flag, refer to photo and insert paper fastener through Flag and Side at ▲. Glue magnetic strip to Mailbox. **For Flag Magnet only:** Follow chart and use required stitches to work Flag Magnet. Glue magnetic strip to wrong side of Magnet.

Designs by Sue Penrod.

NL	COLOR
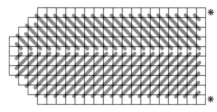 35	blue
41	white
	black embroidery floss

Flag/Flag Magnet (25 x 12 threads)

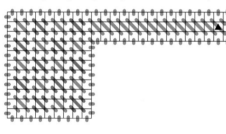

Back (22 x 11 threads)

Bottom (11 x 41 threads)

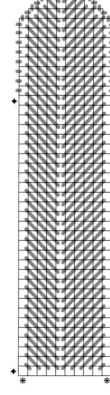

Side (50 x 30 threads)

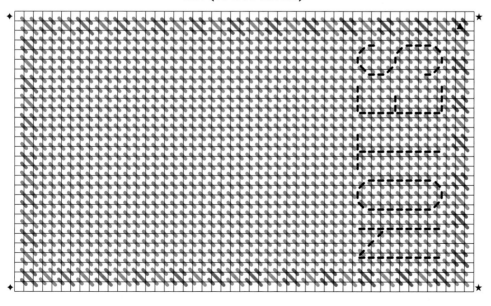

51

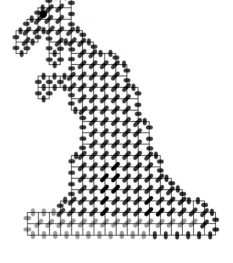

Dandy Dinosaurs

Youngsters will love traveling back through the ages with these prehistoric pals! The colorful dinosaurs will help turn a child's room into "the land before time."

DANDY DINOSAURS

Size: 3 1/8"w x 2 5/8"h each

Supplies: Needloft® Plastic Canvas Yarn or worsted weight yarn (refer to color key), one 10 1/2" x 13 1/2" sheet of 7 mesh plastic canvas, #16 tapestry needle, magnetic strip, and clear-drying craft glue

Stitches Used: Backstitch, French Knot, Overcast Stitch, and Tent Stitch

Instructions: Follow chart and use required stitches to work Dinosaur. Glue magnetic strip to wrong side of Magnet.

Designs by Dick Martin.

Dinosaur #1 (21 x 17 threads)

Dinosaur #2 (19 x 22 threads)

Dinosaur #3 (24 x 16 threads)

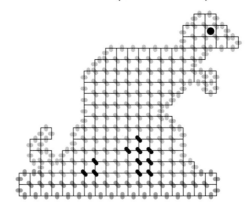

NL	COLOR	
	00	black
	02	red
	23	green
	57	yellow
	64	purple
●	00	black Fr. Knot

Send in the Clowns

What better way to make someone smile than by enlisting the help of a clown! Our jolly twosome is certain to bring loads of cheer wherever they appear.

SEND IN THE CLOWNS

Approx Size: 2"w x 4¼"h each

Supplies: Worsted weight yarn or Needloft® Plastic Canvas Yarn (refer to color key), one 10½" x 13½" sheet of 7 mesh plastic canvas, #16 tapestry needle, magnetic strip, and clear-drying craft glue

Stitches Used: Backstitch, French Knot, Gobelin Stitch, Overcast Stitch, and Tent Stitch

Instructions: Follow charts and use required stitches to work Clown pieces. Glue magnetic strip to wrong side of Magnet. **For Clown A only,** use a dk yellow French Knot and match ▲'s to attach Flower A to Clown A. **For Clown B only,** use pink and match ✱'s to tack Tie B to Clown B.

Designs by Dick Martin.

	NL	COLOR
	00	black
	02	red
	12	dk yellow
	22	lt green
	23	green
	41	white
	57	yellow
	58	orange
	60	lt blue
	62	pink

Clown A (26 x 26 threads)

Flower A (6 x 6 threads)

Clown B Tie (10 x 10 threads)

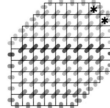

Clown B (26 x 26 threads)

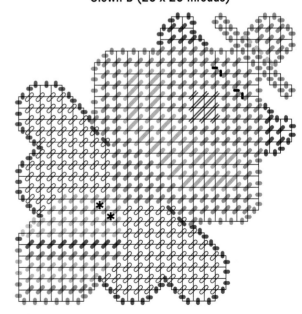

Fresh from the Garden

Fresh-picked vegetables are wholesome, delicious, and good for you, too. We captured some of these garden images with our corn, celery, and carrot magnets. They're so realistic they look good enough to eat!

Corn Front (19 x 19 threads)

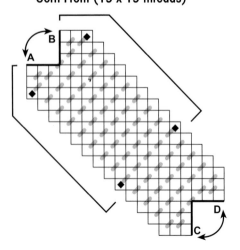

Corn Back (19 x 19 threads)

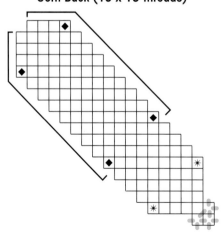

Corn Husk (16 x 16 threads)

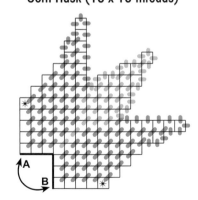

FRESH FROM THE GARDEN

Approx Size: 2"w x 4¼"h each

Supplies: Needloft® Plastic Canvas Yarn or worsted weight yarn (refer to color key), one 10½" x 13½" sheet of 7 mesh plastic canvas, #16 tapestry needle, magnetic strip, and clear-drying craft glue

Stitches Used: Backstitch, Overcast Stitch, and Tent Stitch

Instructions: Follow charts and use required stitches to work Magnet pieces. Glue magnetic strip to wrong side of completed Magnet. **For Carrot only,** match A to B to fold down sections of Front. Use orange to join sections of Front indicated by heavy black lines. Use orange and match ▲'s to join Front to Back along unworked edges. **For Corn only,** match A to B and C to D to fold down sections of Front. Use yellow to join sections of Front indicated by heavy black lines. Match A to B to fold down sections of Husk. Use green to join sections of Husk indicated by heavy black lines. Use yellow and match ◆'s to join Front to Back between brackets. Use green and match ✳'s to join Husk to Front and Back. **For Celery only,** match A to B to fold down sections of Front. Use lt green to join sections of Front indicated by heavy black lines. Use green and match ▲'s to tack Top to Front. Use lt green and match ★'s to join Front to Back along unworked edges.

Designs by Dick Martin.

NL	COLOR
19	yellow* - 5 yds
22	lt green - 8 yds
23	green - 8 yds
52	orange - 3 yds

*Use two strands of yarn.

Carrot Front (15 x 15 threads)

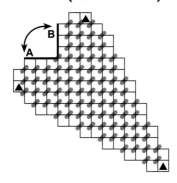

Carrot Back (19 x 19 threads)

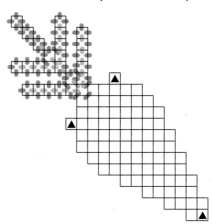

Celery Front (23 x 23 threads)

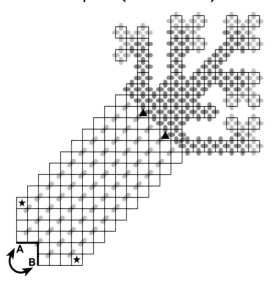

Celery Back (16 x 16 threads)

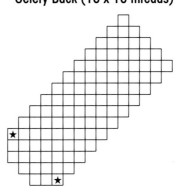

Celery Top (9 x 9 threads)

Great Grapes

Rich in color and shape, this bunch of plump grapes looks like it's just been plucked from the vine!

GREAT GRAPES

Size: 3"w x 3¹/₂"h

Supplies: Needloft® Plastic Canvas Yarn or worsted weight yarn (refer to color key), one 10¹/₂" x 13¹/₂" sheet of 7 mesh plastic canvas, #16 tapestry needle, 5" of green cloth-covered wire, magnetic strip, and clear-drying craft glue

Stitches Used: Backstitch, Overcast Stitch, and Tent Stitch

Instructions: Follow charts and use required stitches to work Grapes pieces. Match A to B to fold down sections of Cluster. Use purple to join sections of Cluster along unworked edges. Cover joining stitches with one dk green stitch. Use purple and match ★'s to tack Cluster to Front. Use purple and match ▲'s to join Front to Back. For tendril, wrap wire around pencil to form a spiral, leaving last ¹/₂" straight. Remove pencil. Insert straight end of wire under stitches on wrong side of Back; glue in place. Glue magnetic strip to wrong side of Magnet.

Design by Dick Martin.

NL	COLOR
14	brown - 1 yd
22	lt green - 2 yds
23	green - 3 yds
46	purple - 8 yds

Front (16 x 16 threads)

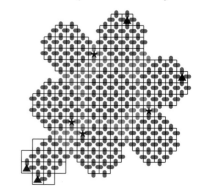

Back (20 x 20 threads)

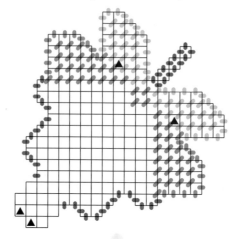

Cluster (10 x 10 threads)

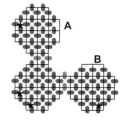

Fun with Trolls

With their delightfully imaginative features, trolls bring a bit of whimsy no matter where they're found. These three are ready for a fun vacation at the beach.

Pick Up Sunscreen!

FUN WITH TROLLS

Approx Size: 2 1/2"w x 5 3/8"h each

Supplies: Needloft® Plastic Canvas Yarn or worsted weight yarn (refer to color key), black embroidery floss, one 10 1/2" x 13 1/2" sheet of 7 mesh plastic canvas, #16 tapestry needle, fine-toothed comb, magnetic strip, and clear-drying craft glue

Stitches Used: Backstitch, French Knot, Fringe, Gobelin Stitch, Overcast Stitch, and Tent Stitch

Instructions: Follow charts and use required stitches to work Magnet pieces. Glue magnetic strip to wrong side of stitched piece. For hair, separate each strand of Fringe into plies. Comb each row of Fringe until the desired look is achieved. **For Girl only,** refer to photo to glue Dress to Girl. **For Troll only,** refer to photo to glue Bow Tie to Troll.

Designs by Dick Martin.

NL	COLOR
43	camel
60	bright blue
61	bright green
62	bright pink
63	bright yellow
64	bright purple
	black embroidery floss*
•	black embroidery floss Fr. Knot*
○ 61	bright green Fringe
○ 62	bright pink Fringe
○ 64	bright purple Fringe

*Use six strands of embroidery floss.

Girl Dress (12 x 10 threads)

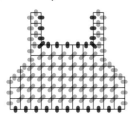

Troll Bow Tie (9 x 9 threads)

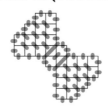

Girl (18 x 27 threads)

Boy (18 x 27 threads)

Troll (16 x 23 threads)

Country Bunnies

Country enthusiasts will love this precious pair of bunnies. The flop-eared friends will warm your heart in a hop, a skip, and a jump!

COUNTRY BUNNIES

Size: 2³/₈"w x 3"h each

Supplies: Worsted weight yarn or Needloft® Plastic Canvas Yarn (refer to color key), six strand grey embroidery floss, one 10¹/₂" x 13¹/₂" sheet of 7 mesh plastic canvas, #16 tapestry needle, magnetic strip, and clear-drying craft glue

Stitches Used: Backstitch, French Knot, Gobelin Stitch, Mosaic Stitch, Overcast Stitch, and Tent Stitch

Instructions: Follow chart and use required stitches to work Bunny. Glue magnetic strip to wrong side of Magnet.

Designs by Sue McElhaney.

Boy Bunny
(15 x 20 threads)

Girl Bunny
(15 x 20 threads)

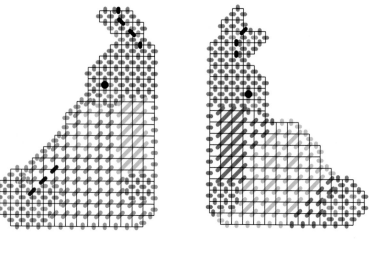

NL	COLOR		NL	COLOR
03	pink		39	cream
05	lt pink		41	white
31	dk blue			grey embroidery floss
33	blue		●	grey embroidery floss Fr. Knot

Farm Folk

Wearing overalls and a workday dress and bonnet, these country folk represent the good old-fashioned virtue of hard work. You'll love their simple look!

FARM FOLK
Approx Size: 2³/₈"w x 4"h each
Supplies: Needloft® Plastic Canvas Yarn or worsted weight yarn (refer to color key), one 10¹/₂" x 13¹/₂" sheet of 7 mesh plastic canvas, #16 tapestry needle, magnetic strip, and clear-drying craft glue
Stitches Used: Backstitch, Gobelin Stitch, Overcast Stitch, and Tent Stitch
Instructions: Follow chart and use required stitches to work Magnet. Glue magnetic strip to wrong side of Magnet. **For Girl only**, thread 6" of blue yarn through canvas at ▲'s; tie in a bow and trim ends.

Designs by Lynette Crawford.

NL	COLOR
✎	00 black
✎	03 red
✎	07 pink
✎	18 tan
✎	33 blue
✎	39 ecru

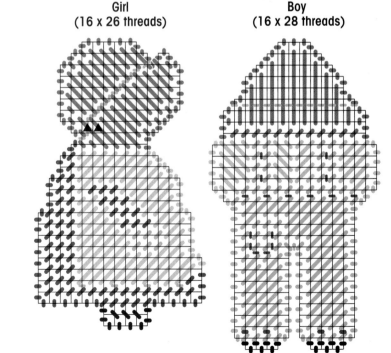

Girl
(16 x 26 threads)

Boy
(16 x 28 threads)

HOT STUFF

Spice up your refrigerator or family message center with these pepper magnets! The south-of-the-border designs can also be strung together to make a chili ristra.

HOT STUFF
Supplies: Worsted weight yarn (refer to photo), one 10½" x 13½" sheet of 7 mesh plastic canvas, #16 tapestry needle, magnetic strip, nylon line, sharp needle, raffia, four rubber bands, and clear-drying craft glue
Stitches Used: Overcast Stitch and Tent Stitch

SINGLE PEPPER MAGNET
Approx Size: 1¼"w x 3"h each
Instructions: Follow chart and use required stitches to work one Pepper. Glue magnetic strip to wrong side of Magnet.

DOUBLE PEPPER MAGNET
Approx Size: 2½"w x 4½"h
Instructions: Follow charts and use required stitches to work one Large Pepper and one Small Pepper, both curving to the left. Cut a small bundle of raffia 6" long. Secure raffia with rubber band ½" from top. Braid raffia for 2½"; secure braid with rubber band. Refer to photo and use nylon line and sharp needle to tack Peppers to braid. Cut a piece of raffia 8" long. Tie raffia in a bow around first rubber band. Glue magnetic strip to wrong side of braid.

PEPPER RISTRA
Approx Size: 5"w x 10½"h
Instructions: Refer to photo, then follow charts and use required stitches to work twelve Large Peppers and ten Small Peppers, half curving to the left and half curving to the right. Cut a bundle of raffia 24" long. Fold raffia in half. Secure raffia with rubber band 2" below fold. Braid raffia to 1" from ends and secure with rubber band. Refer to photo and use nylon line and sharp needle to tack Peppers to braid. Cut a small bundle of raffia 18" long. Tie raffia in a bow around first rubber band.

Designs by Sandra Beavers.

Large Peppers
(10 x 24 threads each)

Small Peppers
(8 x 20 threads each)

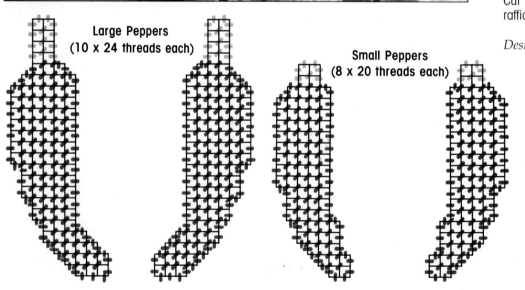

☑ green
☑ pepper color

60

HOME ON THE RANGE

You don't have to live where the deer and the antelope play to enjoy these magnets. They'll bring the flavor of the Old West straight to your home.

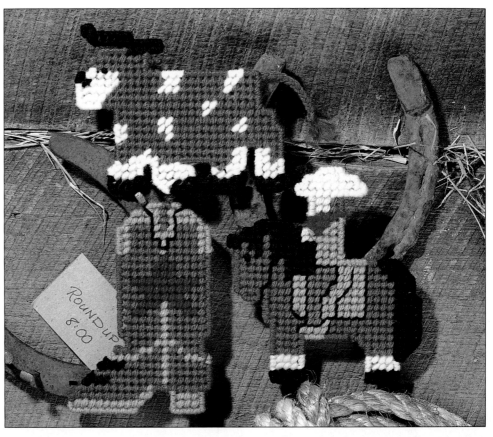

HOME ON THE RANGE

Approx Size: 4"w x 4¼"h each

Supplies: Worsted weight yarn (refer to color key), one 10½" x 13½" sheet of 7 mesh plastic canvas, #16 tapestry needle, magnetic strip, and clear-drying craft glue

Stitches Used: Backstitch, French Knot, Overcast Stitch, and Tent Stitch

Instructions: Follow chart and use required stitches to work magnet. Glue magnetic strip to wrong side of stitched piece. For **Steer only:** Cut three 6" lengths of brown yarn. Knot yarn lengths together close to one end. Refer to photo to thread loose yarn ends under stitches on wrong side of stitched piece. Braid yarn for 2". Knot loose ends of yarn together close to end of braid. Trim ends to ½" and separate yarn into plies.

Designs by Polly Carbonari.

▨ white	▨ lt blue	▨ brown
▨ flesh	▨ blue	▨ black
▨ gold	▨ rust	▨ black 2-ply
▨ red	▨ lt brown	• black 2-ply Fr. Knot

Steer (28 x 22 threads)

Cowboy (27 x 29 threads)

Boot (20 x 28 threads)

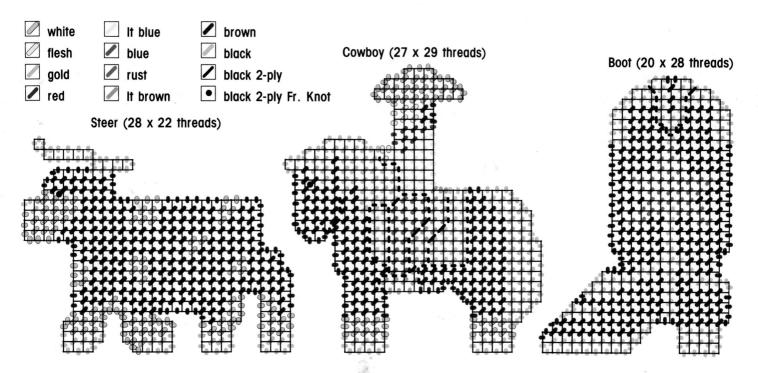

KITCHEN GOURDS

These gourd magnets will add autumn allure to your kitchen. Plant them on the refrigerator, or anywhere you want to leave a trail of fall colors!

KITCHEN GOURDS

Approx Size: 2¼"w x 3⅛"h each

Supplies: Worsted weight yarn (refer to color key), one 10½" x 13½" sheet of 7 mesh plastic canvas, #16 tapestry needle, magnetic strip, and clear-drying craft glue

Stitches Used: Backstitch, French Knot, Overcast Stitch, and Tent Stitch

Instructions: Follow chart and use required stitches to work Magnet. Glue magnetic strip to wrong side of Magnet.

Designs by Virginia Hockenbury.

- ✏ ecru
- ⬜ lt yellow
- ▨ dk yellow
- ▨ orange
- ▨ lt green
- ▨ dk green
- ✏ brown
- ⦿ orange Fr. Knot

Gourd #1 (17 x 18 threads)

Gourd #2 (17 x 28 threads)

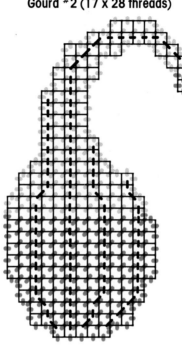

Gourd #3 (12 x 21 threads)

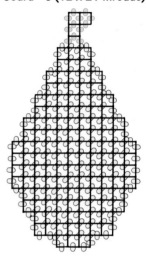

Gourd #4 (17 x 24 threads)

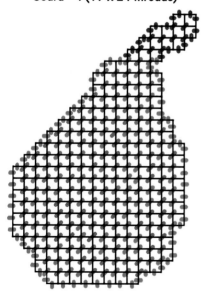

Gourd #5 (13 x 24 threads)

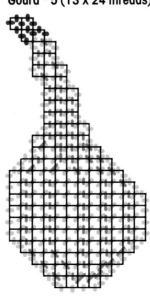

Gourd #6 (17 x 15 threads)

Stitcher's Magnets

Show your fondness for your favorite needlecraft with these creative magnets.

STITCHER'S MAGNETS

Approx Size: 2³/₈"w x 3¹/₂"h each

Supplies: Needloft® Plastic Canvas Yarn or worsted weight yarn (refer to color key), metallic silver yarn, black embroidery floss, one 10¹/₂" x 13¹/₂" sheet of 7 mesh plastic canvas, #16 tapestry needle, one 1" x 5" piece of cardboard, magnetic strip, and clear-drying craft glue

Stitches Used: Backstitch, Gobelin Stitch, Overcast Stitch, and Tent Stitch

Instructions: Use six strands of embroidery floss. Follow charts and use required stitches to work Magnet pieces. Glue magnetic strip to wrong side of Magnet.

For PC Basket only: For PC Sheets, cut one piece of plastic canvas 6 x 9 threads and one piece of plastic canvas 8 x 11 threads. For yarn skein, wrap 14" of red yarn around short end of cardboard. Slide yarn bundle from cardboard. Wrap and glue 6" of red yarn around center of yarn skein. Refer to photo to glue yarn skein and PC Needle to PC Basket. Glue PC Sheets to wrong side of PC Basket.

Designs by Jocelyn Sass.

NL	COLOR		NL	COLOR
01	red - 2 yds		39	eggshell - 4 yds
12	pumpkin - 2 yds			metallic silver - 1 yd
37	silver - 2 yds			black embroidery floss - 1 yd

PC Needle
(10 x 10 threads)

Scissors (16 x 27 threads)

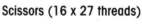

PC Basket (19 x 22 threads)

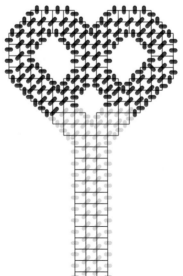

Apple Appeal

Looking good enough to eat, this cute apple magnet appears to have been plucked fresh from the tree! The appealing red fruit, stitched on 14 mesh plastic canvas, is sure to become a keepsake.

APPLE APPEAL

Size: 2"w x 2³/₈"h

Supplies: DMC embroidery floss (refer to color key), one 8" x 11" sheet of 14 mesh plastic canvas, #24 tapestry needle, magnetic strip, and clear-drying craft glue

Stitches Used: Backstitch, Gobelin Stitch, Overcast Stitch, and Tent Stitch

Instructions: Follow chart and use required stitches to work Magnet, using six strands of embroidery floss. Glue magnetic strip to wrong side of Magnet.

Design by Kathleen J. Fischer.

	DMC	COLOR
	blanc	white - 1 yd
	321	red - 8 yds
	433	brown - 1 yd
	699	dk green - 1 yd
	701	green - 2 yds

Apple (28 x 33 threads)

Happy Easter!

These sweet images of Easter will add charm to the holiday. Whether you stitch one or all four, the collection will get you in the spirit of spring, too!

HAPPY EASTER!

Approx Size: 3¾"w x 4¾"h each

Supplies: Worsted weight yarn or Needloft® Plastic Canvas Yarn (refer to color key), one 10½" x 13½" sheet of 7 mesh plastic canvas, #16 tapestry needle, magnetic strip, and clear-drying craft glue

Stitches Used: Backstitch, French Knot, Gobelin Stitch, Lazy Daisy Stitch, Overcast Stitch, and Tent Stitch

Instructions: Follow chart and use required stitches to work Magnet. Glue magnetic strip to wrong side of completed Magnet. **For Bunny only**, Thread 8" of pink yarn through canvas at ◆. Tie yarn in a bow and trim ends. **For Duck only**, Tie 8" of green yarn in a knot around neck and trim ends. **For Rabbit only**, Tie 8" of yellow yarn in a bow around neck and trim ends.

Designs by Dick Martin.

NL	COLOR
07	pink
12	orange
13	brown
20	lt yellow
22	green
41	white

NL	COLOR
45	lavender
51	aqua
57	yellow
07	pink Fr. Knot
45	lavender Fr. Knot
57	yellow Fr. Knot

NL	COLOR
	orange Fr. Knot*
	black Fr. Knot**
41	white Lazy Daisy

*Use 2-ply yarn.

**Use 1-ply yarn.

Chick (19 x 19 threads)

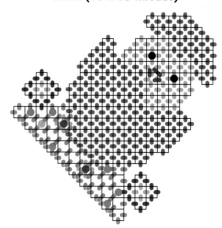

Rabbit (21 x 17 threads)

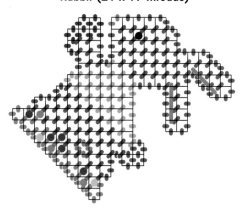

Bunny (19 x 19 threads)

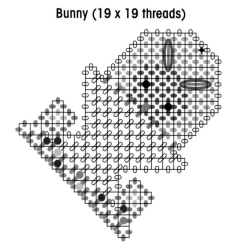

Duck (20 x 17 threads)

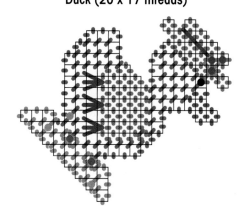

Very Victorian

This frilly feminine set will appeal to those who love the elegance of the Victorian Era. The magnets will add a soft touch wherever you display them.

Garden Brunch
Sunday
1:00

VERY VICTORIAN

Approx Size: 3½"w x 3"h each

Supplies: Needloft® Plastic Canvas Yarn or worsted weight yarn (refer to color key), embroidery floss (refer to color key), metallic gold yarn, one 10½" x 13½" sheet of 7 mesh plastic canvas, #16 tapestry needle, 8" of ¼"w blue satin ribbon, magnetic strip, and clear-drying craft glue

Stitches Used: Backstitch, Cross Stitch, French Knot, Fringe, Gobelin Stitch, Lazy Daisy Stitch, Overcast Stitch, Smooth Spider Web Stitch, and Tent Stitch

Instructions: Follow chart and use required stitches to work Magnet. Glue magnetic strip to wrong side of Magnet. **For Swan**, tie ribbon in a bow and trim ends. Refer to photo to glue bow to stitched piece. **For Lamp**, refer to photo to trim Fringe.

Cream Pitcher and Sugar Bowl designs by Lora Neal.
Fan design by Virginia Hockenbury.
Lamp design by Marie Leppek.
Swan design by Kay Stuckey.

NL	COLOR
05	lavender - 2 yds
07	pink - 2 yds
12	pumpkin - 1 yd
19	straw - 1 yd
39	eggshell - 16 yds
53	green - 7 yds
	black embroidery floss* - 1 yd
	grey embroidery floss* - 2 yds
	metallic gold - 4 yds
00	black Fr. Knot - 1 yd
05	lavender Fr. Knot
	green embroidery floss Lazy Daisy* - 5 yds
	ecru embroidery floss Fringe* - 4 yds

*Use 6 strands of embroidery floss.

Lamp (18 x 24 threads)

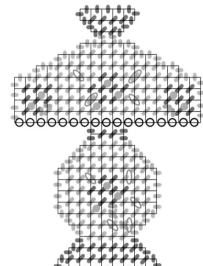

Fan (28 x 17 threads)

Sugar Bowl (23 x 19 threads)

Cream Pitcher (16 x 17 threads)

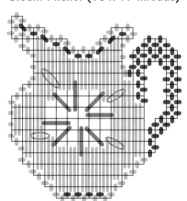

Swan (31 x 21 threads)

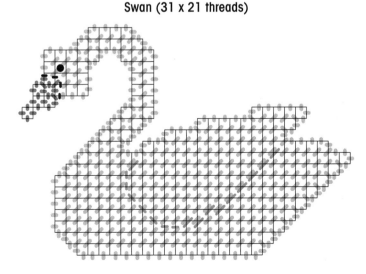

Candy Cane Santa

Express your love for Christmas (and Santa Claus!) with this festive heart. Stitched on 7 mesh plastic canvas, the sweet magnet works up in a jiffy. You'll want to keep it for years to come!

CANDY CANE SANTA

Size: 4"w x 4"h

Supplies: Needloft® Plastic Canvas Yarn or worsted weight yarn (refer to color key), embroidery floss (refer to color key), one 10¹/₂" x 13¹/₂" sheet of 7 mesh plastic canvas, #16 tapestry needle, magnetic strip, and clear-drying craft glue

Stitches Used: Backstitch, Cross Stitch, French Knot, Gobelin Stitch, Overcast Stitch, and Tent Stitch

Instructions: Use six strands of embroidery floss for French Knots. Follow chart and use required stitches to work Magnet. Glue magnetic strip to wrong side of Magnet.

Design by Dick Martin.

Candy Cane Heart
(27 x 27 threads)

NL	COLOR
02	Christmas red
07	pink
27	holly
41	white
56	flesh tone
	red embroidery floss Fr. Knot
	black embroidery floss Fr. Knot

"Purr-fect" Pair

Cat fanciers will find these magnets simply "purr-fect"! The cat and mouse set will provide a whimsical touch for any room.

"PURR-FECT" PAIR

Approx Size: 2³/₄"w x 3¹/₄"h each

Supplies: Needloft® Plastic Canvas Yarn or worsted weight yarn (refer to color key), six strand black embroidery floss, one 10¹/₂" x 13¹/₂" sheet of 7 mesh plastic canvas, #16 tapestry needle, magnetic strip, and clear-drying craft glue

Stitches Used: Backstitch, French Knot, Overcast Stitch, and Tent Stitch

Instructions: Follow chart and use required stitches to work Magnet. Glue magnetic strip to wrong side of Magnet.

Mouse (19 x 18 threads)

Cat (19 x 27 threads)

NL	COLOR		NL	COLOR		NL	COLOR
	05 pink			11 lt tan			41 white
	07 lt pink			12 tan			black embroidery floss
	09 dk tan			33 blue			05 pink Fr. Knot

Country Favorites

Show your pride in the U.S.A. with these symbols of Americana — a patriotic heart, a basket of apples, and a delicious pie. The magnets are fun for the Fourth of July or all year round!

COUNTRY FAVORITES
Approx Size: 3"w x 3"h each
Supplies: Needloft® Plastic Canvas Yarn or worsted weight yarn (refer to color key), black embroidery floss, one 10½" x 13½" sheet of 7 mesh plastic canvas, #16 tapestry needle, magnetic strip, and clear-drying craft glue
Stitches Used: Backstitch, French Knot, Overcast Stitch, and Tent Stitch
Instructions: Use six strands of embroidery floss. Follow chart and use required stitches to work Magnet. Glue magnetic strip to wrong side of Magnet.

Designs by Lorraine Birmingham.

NL	COLOR
01	red
13	maple
14	cinnamon
16	sandstone
29	forest
32	royal
35	sail blue
41	white
42	crimson
43	camel
48	royal dark
	black embroidery floss
41	white Fr. Knot

Patriotic Heart (23 x 20 threads)

Apple Basket (25 x 19 threads)

Pie (27 x 13 threads)

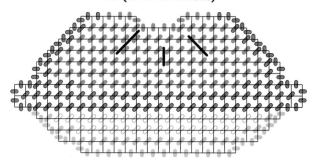

Quick & Colorful

Perk up your kitchen with these quick-to-stitch projects. The colorful images will be a bright spot in anyone's day.

QUICK & COLORFUL

Approx Size: 2⁵/₈"w x 2³/₈"h each

Supplies: Needloft® Plastic Canvas Yarn or worsted weight yarn (refer to color key), six strand black embroidery floss, one 10¹/₂" x 13¹/₂" sheet of 7 mesh plastic canvas, #16 tapestry needle, magnetic strip, and clear-drying craft glue

Stitches Used: Backstitch, Cross Stitch, French Knot, Gobelin Stitch, Overcast Stitch, and Tent Stitch

Instructions: Follow chart and use required stitches to work Magnet. Glue magnetic strip to wrong side of Magnet.

NL	COLOR
02	red
14	brown
22	lt green
23	green
32	dk blue

NL	COLOR
35	blue
38	grey
41	white
43	tan
46	purple

NL	COLOR
52	orange
57	yellow
32	dk blue Fr. Knot
	black embroidery floss

Rainbow (17 x 14 threads)

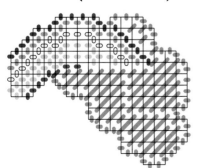

House (20 x 17 threads)

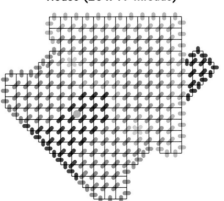

Balloon (17 x 17 threads)

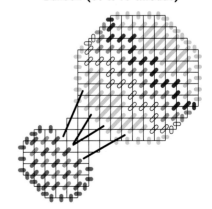

Save Our Wildlife

The protection of endangered species is important to people around the world, and these magnets reflect that concern. The three timely designs will let you share your sentiments with others.

SAVE OUR WILDLIFE

Approx Size: 3½"w x 3½"h each

Supplies: Needloft® Plastic Canvas Yarn or worsted weight yarn (refer to color key), embroidery floss (refer to color key), one 10½" x 13½" sheet of 7 mesh plastic canvas, #16 tapestry needle, magnetic strip, and clear-drying craft glue

Stitches Used: Backstitch, French Knot, and Overcast Stitch

Instructions: Use six strands of embroidery floss for French Knots. Follow chart and use required stitches to work Magnet. Glue magnetic strip to wrong side of completed Magnet.

Designs by Dick Martin.

NL	COLOR		NL	COLOR
00	black		46	purple
10	sundown		48	royal dark
16	sandstone		54	turquiose
22	lime			white embroidery floss Fr. Knot
27	holly			dk green embroidery floss Fr. Knot
41	white			black embroidery floss Fr. Knot

"Save Me" Panda (18 x 23 threads)

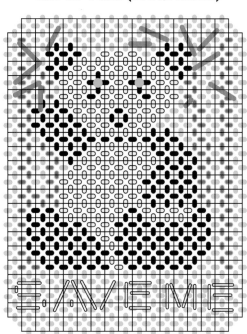

Save The Whale (22 x 17 threads)

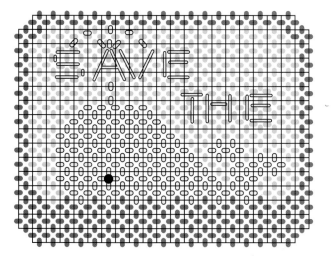

"Save Me" Tiger (24 x 18 threads)

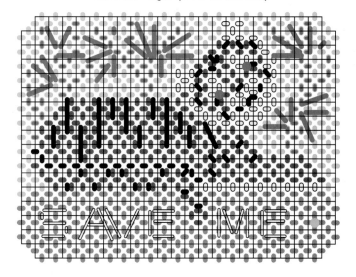

Show-Off Magnets

Show off your favorite little student's best work by hanging it with these fun magnets. Whether you're showcasing a perfect grade or a delightful drawing, this pair of holders will get the job done.

SHOW-OFF MAGNETS
Approx Size: 2³/₄"w x 2¹/₂"h each
Supplies: Worsted weight yarn or Needloft® Plastic Canvas Yarn (refer to color key), one 10¹/₂" x 13¹/₂" sheet of 7 mesh plastic canvas, #16 tapestry needle, magnetic strip, and clear-drying craft glue
Stitches Used: Backstitch, Overcast Stitch, and Tent Stitch
Instructions: Follow chart and use required stitches to work Magnet. Glue magnetic strip to wrong side of Magnet.

Designs by Dick Martin.

NL	COLOR		NL	COLOR
11	gold - 3 yds		60	blue - 3 yds
23	green - 2 yds		62	pink - 3 yds
58	orange - 2 yds			

#1 (17 x 17 threads)

A+ (19 x 16 threads)

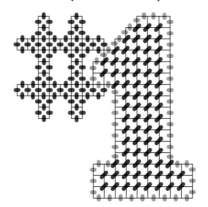

In the Moo-d

These cow magnets are "udderly a-moo-sing"! Quick to stitch, they'll bring smiles to all who see them.

IN THE MOO-D

Size: 2³/₈"w x 2¹/₂"h

Supplies: Worsted weight yarn or Needloft® Plastic Canvas Yarn (refer to color key), one 10¹/₂" x 13¹/₂" sheet of 7 mesh plastic canvas, #16 tapestry needle, one ¹/₂" cowbell, 8" of ¹/₈"w pink satin ribbon, magnetic strip, and clear-drying craft glue

Stitches Used: Backstitch, Overcast Stitch, and Tent Stitch

Instructions: Follow chart and use required stitches to work Magnet. Refer to photo and use pink to tack cowbell to wrong side of Magnet. Tie ribbon in a bow and trim ends. Refer to photo to glue bow to Magnet. Glue magnetic strip to wrong side of Magnet.

Design by Vicki M. Rohner.

Cow (16 x 18 threads)

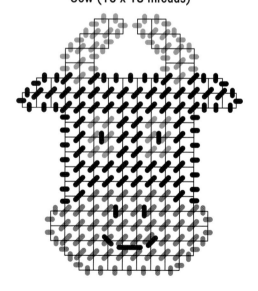

NL	COLOR	
✎	00	black - 2 yds
✎	07	pink - 2 yds
✎	41	white - 1 yd

Pretty Poinsettia

This lovely poinsettia symbolizes the beauty of Christmas. Known as the Flower of the Holy Night, the striking bloom has long been a favorite decoration during the Yuletide season.

PRETTY POINSETTIA

Size: 3³/₄"w x 3⁵/₈"h

Supplies: Worsted weight yarn or Needloft® Plastic Canvas Yarn (refer to color key), one 10¹/₂" x 13¹/₂" sheet of 7 mesh plastic canvas, #16 tapestry needle, magnetic strip, and clear-drying craft glue

Stitches Used: French Knot, Overcast Stitch, and Tent Stitch

Instructions: Follow chart and use required stitches to work Poinsettia. Glue magnetic strip to wrong side of Magnet.

Design by Joan E. Ray.

Poinsettia (26 x 25 threads)

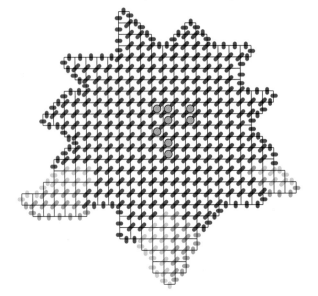

NL	COLOR
02	red - 4 yds
27	lt green - 2 yds
29	green - 2 yds
42	dk red - 4 yds
19	yellow Fr. Knot - 1 yd

Just for Teacher

The clever designs in this back-to-school set definitely merit an A+! They'll make great gifts for a child's favorite teacher.

JUST FOR TEACHER

Approx Size: 3³⁄₈"w x 1³⁄₄"h each

Supplies: Needloft® Plastic Canvas Yarn or worsted weight yarn (refer to color key), metallic gold braid, one 10¹⁄₂" x 13¹⁄₂" sheet of 7 mesh plastic canvas, #16 tapestry needle, magnetic strip, and clear-drying craft glue

Stitches Used: Backstitch, Gobelin Stitch, Overcast Stitch, and Tent Stitch

Instructions: Follow chart and use required stitches to work Magnet. Glue magnetic strip to wrong side of Magnet.

Designs by Dick Martin.

NL	COLOR		NL	COLOR		NL	COLOR
00	black		23	fern		57	yellow
02	red		41	white			metallic gold braid
07	pink		43	camel			

Pencil (18 x 18 threads)

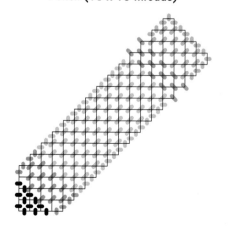

Apple (15 x 12 threads)

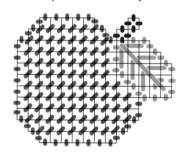

Chalkboard (12 x 16 threads)

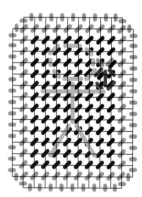

Ruler (41 x 7 threads)

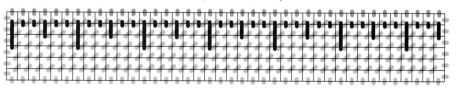

Halloween Hobgoblins

This host of haunting magnets will make great little gifts for your favorite Halloween tricksters. The cute characters can also be used to create a playful wreath.

HALLOWEEN HOBGOBLINS

Stitches Used: Backstitch, Cross Stitch, French Knot, Gobelin Stitch, Overcast Stitch, and Tent Stitch

MAGNETS

Supplies For One Of Each Magnet: Worsted weight yarn or Needloft® Plastic Canvas Yarn (refer to color key), one 10 1/2" x 13 1/2" sheet of 7 mesh plastic canvas, #16 tapestry needle, four 7mm moving eyes, magnetic strip, and clear-drying craft glue

Instructions: Follow chart and use required stitches to work magnet. Glue magnetic strip to wrong side of stitched piece. **For Bat only:** Refer to photo to glue moving eyes to Bat. **For Monster only:** Refer to photo to glue moving eyes to Monster. Refer to photo to thread 5" of orange yarn through canvas at ✚ 's. Tie yarn in a knot close to canvas. Tie each yarn end in a knot 1" from canvas and trim ends.

WREATH

Size: 16" diameter
Supplies: Worsted weight yarn or Needloft® Plastic Canvas Yarn (refer to color key), one 10 1/2" x 13 1/2" sheet of 7 mesh plastic canvas, #16 tapestry needle, two 7mm moving eyes, 2 yards of 1 1/2"w craft ribbon, 16" diameter grapevine wreath, 30" long grapevine leaf stem, 10"w x 4"h wooden picket fence, and clear-drying craft glue
Instructions: Follow charts and use required stitches to work one Ghost, one Bat, two Small Pumpkins, and three Large Pumpkins. Refer to photo to glue moving eyes to Bat. Refer to photo to arrange and glue grapevine leaf stem and picket fence to wreath. Use craft ribbon to make a multi-loop bow. Glue bow to wreath and trim ends. Glue stitched pieces to wreath.

Designs by Virginia Hockenbury.

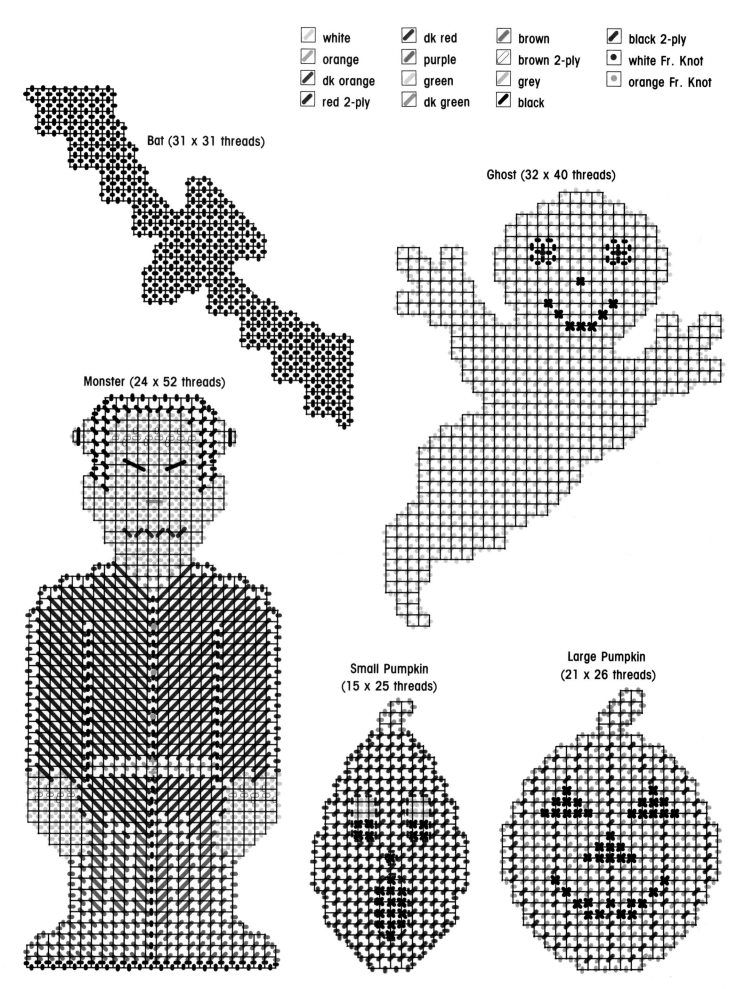

	white		dk red		brown		black 2-ply
	orange		purple		brown 2-ply	•	white Fr. Knot
	dk orange		green		grey		orange Fr. Knot
	red 2-ply		dk green		black		

Bat (31 x 31 threads)

Ghost (32 x 40 threads)

Monster (24 x 52 threads)

Small Pumpkin
(15 x 25 threads)

Large Pumpkin
(21 x 26 threads)

RASCALLY RACCOON

This little scamp will steal your heart with his mischievous ways! Our rascally raccoon is perfect for hanging notes or just hanging around.

RASCALLY RACCOON
Size: 4"w x 3½"h
Supplies: Worsted weight yarn or Needloft® Plastic Canvas Yarn (refer to color key), one 10½" x 13½" sheet of 7 mesh plastic canvas, #16 tapestry needle, two 5mm brown animal eyes, magnetic strip, and clear-drying craft glue
Stitches Used: Backstitch, Cross Stitch, Gobelin Stitch, Overcast Stitch, and Tent Stitch
Instructions: Follow charts and use required stitches to work Magnet pieces. Refer to photo to glue eyes to Head. Glue Head and Paws to Watermelon. Glue magnetic strip to wrong side of stitched piece.

Design by Darla J. Fanton.

NL	COLOR
✏	00 black - 2 yds
✏	22 green - 1 yd
✏	27 dk green - 1 yd
✏	37 grey - 2 yds

NL	COLOR
✏	38 dk grey - 1 yd
✏	39 ecru - 2 yds
✏	41 white - 1 yd
✏	55 pink - 3 yds

Paw
(4 x 4 threads)
(Work 2)

Watermelon (20 x 20 threads)

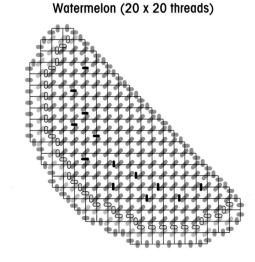

Head (17 x 17 threads)

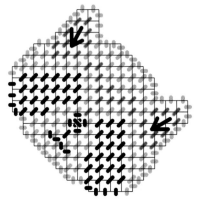

82

Sunny Bloom

The sunflower's brilliance is forever captured in this bright magnet. Its vibrant colors will be reminders of the showy summer flower all year long.

SUNNY BLOOM

Size: 4³/₈"w x 4³/₈"h

Supplies: Needloft® Plastic Canvas Yarn or worsted weight yarn (refer to color key), black embroidery floss, one 10¹/₂" x 13¹/₂" sheet of 7 mesh plastic canvas, #16 tapestry needle, magnetic strip, and clear-drying craft glue

Stitches Used: Backstitch, Gobelin Stitch, Overcast Stitch, and Tent Stitch

Instructions: Use six strands of embroidery floss. Follow charts and use required stitches to work Magnet pieces, leaving stitches in shaded area unworked. Refer to photo to place Base on Petals. Work stitches in shaded area to join Base to Petals. Use yellow and match ★'s to tack Petals to right side of Base. Glue magnetic strip to wrong side of Magnet.

Design by Dick Martin.

NL	COLOR
00	black
22	lime

NL	COLOR
23	fern
43	camel

NL	COLOR
57	yellow
	black embroidery floss

Petals (13 x 13 threads)

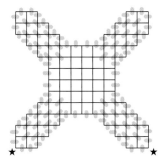

Base (21 x 21 threads)

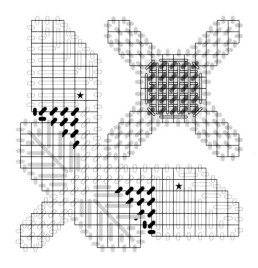

Canada Geese

As winter approaches, geese begin their annual flight in search of warmer weather. Frozen in time, these majestic Canada geese make a striking pair for a favorite sportsman.

CANADA GEESE

Approx Size: 2³/₄"w x 2³/₄"h each

Supplies: Worsted weight yarn or Needloft® Plastic Canvas Yarn (refer to color key), six strand brown embroidery floss, one 10¹/₂" x 13¹/₂" sheet of 7 mesh plastic canvas, #16 tapestry needle, magnetic strip, and clear-drying craft glue

Stitches Used: Backstitch, Gobelin Stitch, and Overcast Stitch

Instructions: Follow chart and use required stitches to work Magnet. Glue magnetic strip to wrong side of Magnet.

Designs by Dick Martin.

Goose #1 (21 x 16 threads)

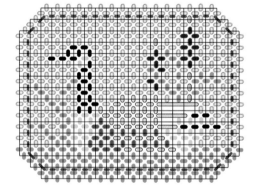

Goose #2 (16 x 21 threads)

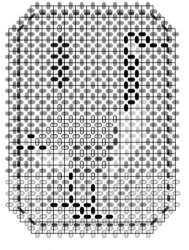

NL	COLOR		NL	COLOR		NL	COLOR		NL	COLOR		NL	COLOR
00	black		15	brown		29	green		39	ecru		43	taupe
14	dk taupe		17	gold		38	grey		40	tan			brown embroidery floss

For Mom & Dad

These cute teddy bear magnets are great for reminding parents just how special they are. Your mom and dad will be proud to display them on the fridge or at work.

FOR MOM & DAD

Size: 2¹/₂"w x 3"h

Supplies: Needloft® Plastic Canvas Yarn or worsted weight yarn (refer to color key), one 10¹/₂" x 13¹/₂" sheet of clear 7 mesh plastic canvas, one 10¹/₂" x 13¹/₂" sheet of pink 7 mesh plastic canvas, one 10¹/₂" x 13¹/₂" sheet of blue 7 mesh plastic canvas, #16 tapestry needle, paper, pen, magnetic strip, and clear-drying craft glue

Stitches Used: Cross Stitch, French Knot, Gobelin Stitch, Overcast Stitch, and Tent Stitch

Instructions: Follow charts and use required stitches to work Bear Body and Bear Paws. Use brown and match ▲'s to join Bear Paws to Bear Body. **For Mom Bear,** cut Frame Front and Frame Back from pink plastic canvas. **For Dad Bear,** cut Frame Front and Frame Back from blue plastic canvas. Cut paper the same size as plastic canvas. Write message on paper. Glue paper between Front and Back. Glue Frame to Bear's paws. Glue magnetic strip to wrong side of Magnet.

Designs by Melanna Rosenthal and Joan Bartling.

NL	COLOR		NL	COLOR
✎	00 black		✎	16 sandstone
✎	15 brown		●	00 black Fr. Knot

Frame Front
(10 x 9 threads)

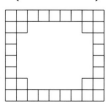

Frame Back
(10 x 9 threads)

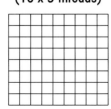

Bear Paws
(6 x 6 threads)
(Work 2)

Bear Body (21 x 21 threads)

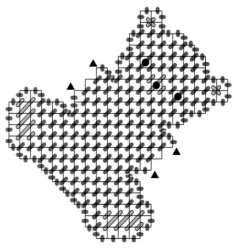

85

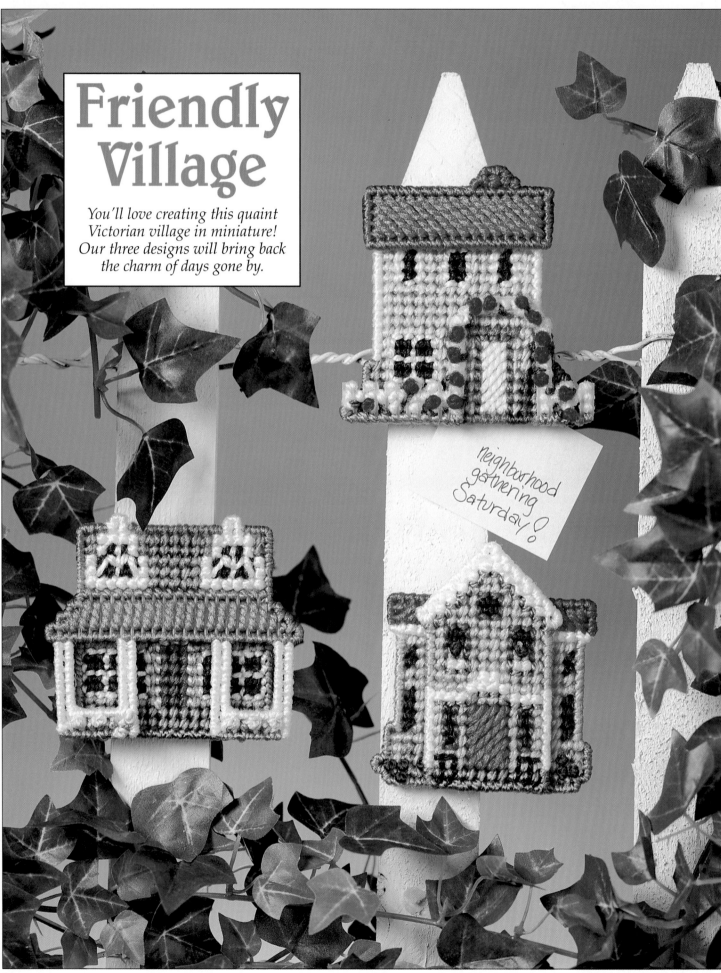

Friendly Village

You'll love creating this quaint Victorian village in miniature! Our three designs will bring back the charm of days gone by.

neighborhood gathering Saturday!

FRIENDLY VILLAGE

Approx Size: 3½"w x 3⅜"h each
Supplies: Needloft® Plastic Canvas Yarn or worsted weight yarn (refer to color key), two yards of red yarn, one 10½" x 13½" sheet of 7 mesh plastic canvas, #16 tapestry needle, magnetic strip, and clear-drying craft glue
Stitches Used: Backstitch, Cross Stitch, French Knot, Gobelin Stitch, Overcast Stitch, and Tent Stitch
Instructions: Follow charts and use required stitches to work Magnet pieces. Glue magnetic strip to wrong side of completed Magnet. **For House A,** use lt brown and match ★'s to join Roof to House A. Use lt green and match ◆'s to join Fence to House A along unworked bottom edges. Refer to photo and use red to work French Knots through Fence and House A. **For House B,** use tan and match ▲'s to join Porch to House B between ▲'s. Glue bottom edge of Porch to House B. **For House C,** match ◆'s to glue Front to Back.

Designs by Kooler Design Studio.

NL	COLOR
10	rust
11	orange
13	lt brown
15	brown
19	lt yellow
23	lt green
27	green
33	blue
34	lt blue
38	grey
41	white
43	tan
47	peach
19	lt yellow Fr. Knot
45	purple Fr. Knot

House A (22 x 24 threads)

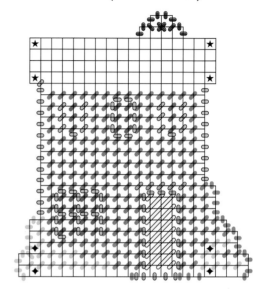

House A Roof (18 x 6 threads)

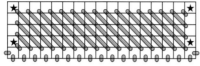

House A Fence (24 x 12 threads)

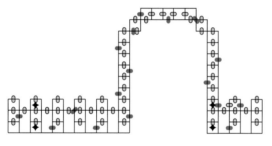

House B (19 x 21 threads)

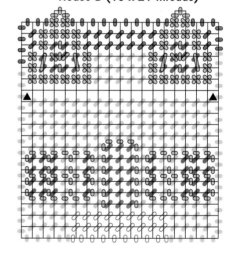

House C Front (14 x 23 threads)

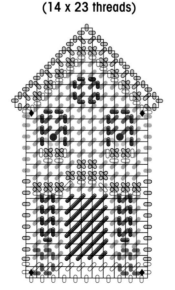

House B Porch (25 x 14 threads)

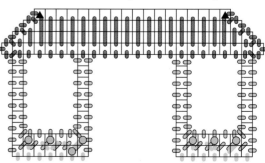

House C Back (21 x 18 threads)

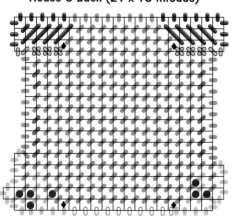

LITTLE LAMBS

Mary had a little lamb but it wasn't as cute as these! Wearing wildflower garlands, they're ready to celebrate spring.

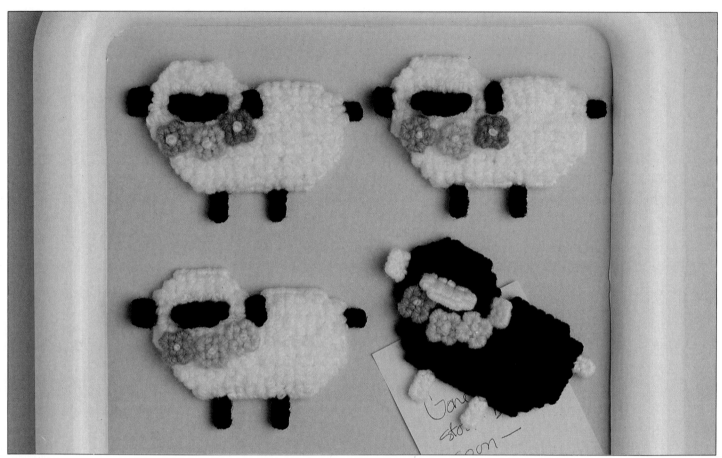

LITTLE LAMBS
Size: 3¹/₂"w x 2³/₄"h
Supplies: Worsted weight yarn (refer to color key), one 10¹/₂" x 13¹/₂" sheet of 7 mesh plastic canvas, #16 tapestry needle, magnetic strip, and clear-drying craft glue
Stitches Used: Backstitch, French Knot, Gobelin Stitch, Overcast Stitch, and Turkey Loop
Instructions: Follow chart and use required stitches to work Lamb pieces. (**Note:** For black lamb, reverse black and white in color key and instructions.) Match ♦'s and use black to join Face to Body along unworked threads. Match ★'s and △'s and use black to join each Ear to Body. Refer to photo to glue Flowers to Body. Glue magnetic strip to wrong side of Magnet.

⬜⟋	white
⬜⟋	flower color
⬛⟋	black
⊙	yellow 2-ply Fr. Knot
○	white Turkey Loop Stitch

Flower (4 x 4 threads)
(Work 3)

Face (6 x 4 threads)

Body (22 x 16 threads)

Ear (3 x 3 threads)
(Work 2)

Lovely Hyacinths

Known for its beauty and fragrance, the hyacinth has long been a favorite of gardeners. Now you can keep the pretty flowers abloom year-round when you stitch this lovely little magnet in a variety of colors.

LOVELY HYACINTHS

Size: 2"w x 4³/₈"h

Supplies: Worsted weight yarn or Needloft® Plastic Canvas Yarn (refer to color key), one 10¹/₂" x 13¹/₂" sheet of 7 mesh plastic canvas, #16 tapestry needle, magnetic strip, and clear-drying craft glue

Stitches Used: Backstitch, Double French Knot, Gobelin Stitch, Overcast Stitch, and Tent Stitch

Instructions: Follow charts and use required stitches to work Magnet pieces. Refer to photo to glue Leaves to wrong side of Hyacinth. Glue magnetic strip to wrong side of stitched piece.

Design by Teal Lee Elliott.

NL	COLOR
⬜	05 rose
⬜	27 holly
⬜	flower color
⬜	flower color Double Fr. Knot

Leaf A
(5 x 8 threads)

Leaf B
(5 x 8 threads)

Hyacinth
(13 x 29 threads)

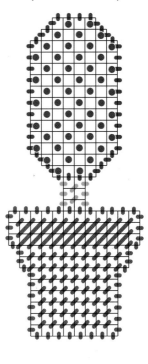

All Aflutter

Grace your home with the beauty of butterflies. Our five pretty magnets look so lifelike you might mistake them for the real thing!

Butterfly
Collection
due Friday

ALL AFLUTTER

Approx Size: 2³/₈"w x 2³/₈"h each

Supplies: Needloft® Plastic Canvas Yarn or worsted weight yarn (refer to color key), six strand embroidery floss (refer to color key), one 10¹/₂" x 13¹/₂" sheet of 7 mesh plastic canvas, #16 tapestry needle, black cloth-covered wire, magnetic strip, and clear-drying craft glue

Stitches Used: Backstitch, French Knot, Overcast Stitch, and Tent Stitch

Instructions: Follow chart and use required stitches to work Magnet, leaving stitches in shaded area unworked. Cut a 2¹/₂" length of wire and shape into a "V". Place bottom of "V" at ♦ on wrong side of Magnet. Work stitches in shaded area, catching wire in stitching. Glue magnetic strip to wrong side of Magnet.

Designs by Dick Martin.

NL	COLOR
00	black
11	lt gold
14	brown
40	beige
41	white
52	orange
54	blue
57	yellow
	cream embroidery floss
	lt gold embroidery floss
●	black embroidery floss Fr. Knot

Butterfly #1 (20 x 20 threads)

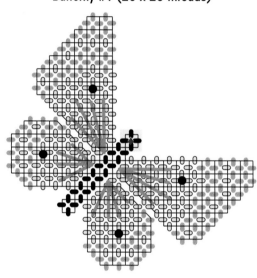

Butterfly #2 (19 x 19 threads)

Butterfly #3 (17 x 17 threads)

Butterfly #4 (14 x 14 threads)

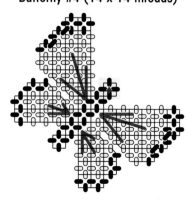

Butterfly #5 (12 x 12 threads)

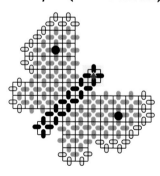

GENERAL INSTRUCTIONS

SELECTING PLASTIC CANVAS

Plastic canvas is a molded, nonwoven canvas made from clear or colored plastic. The canvas consists of "threads" and "holes." The threads aren't actually "threads" since the canvas is nonwoven, but it seems to be an accurate description of the straight lines of the canvas. The holes, as you would expect, are the spaces between the threads. The threads are often referred to in the project instructions, especially when cutting out plastic canvas pieces. The instructions for stitches will always refer to holes when explaining where to place your needle to make a stitch.

TYPES OF CANVAS

The main difference between types of plastic canvas is the mesh size. Mesh size refers to the number of holes in one inch of canvas. The most common mesh sizes are 5 mesh, 7 mesh, 10 mesh, and 14 mesh. Five mesh means that there are 5 holes in every inch of canvas. Likewise, there are 7 holes in every inch of 7 mesh canvas, 10 holes in every inch of 10 mesh canvas, and 14 holes in every inch of 14 mesh canvas. Seven mesh canvas is the most popular size for the majority of projects.

Your project supply list will tell you what size mesh you need to buy. Be sure to use the mesh size the project instructions recommend. If your project calls for 7 mesh canvas and you use 10 mesh, your finished project will be much smaller than expected.

Most plastic canvas is made from clear plastic, but colored canvas is also available. Colored plastic is ideal when you don't want to stitch the entire background.

AMOUNT OF CANVAS

The project supply list usually tells you how much canvas you will need to complete the project. When buying your canvas, remember that several different manufacturers produce plastic canvas. Therefore, there are often slight variations in canvas, such as different thicknesses of threads or a small difference in mesh size. Because of these variations, try to buy enough canvas for your entire project at the same time and place. As a general rule, it is always better to buy too much canvas and have leftovers than to run out of canvas before you finish your project. By buying a little extra canvas, you not only allow for mistakes, but have extra canvas for practicing your stitches.

SELECTING YARN

You're probably thinking, "How do I select my yarn from the thousands of choices available?" Well, we have a few hints to help you choose the perfect yarns for your project and your budget.

To help you select colors for your projects, we have included numbers for Needloft® Plastic Canvas Yarn in our color keys. The headings in the color key are for Needloft® Yarn (**NL**) and the descriptive color name (**COLOR**). Needloft® Yarn is 100% nylon and is suitable only for 7 mesh plastic canvas.

Needloft® Yarn will not easily separate. When the instructions call for 2-ply or 1-ply yarn, we recommend that you substitute with six strands of embroidery floss.

TYPES OF YARN

The first question to ask when choosing yarn is, "How will my project be used?" If your finished project will be handled or used a lot, such as a magnet, you will want to use a durable, washable yarn. We highly recommend acrylic or nylon yarn for plastic canvas. It can be washed repeatedly and holds up well to frequent usage and handling.

Cost may also be a factor in your yarn selection. There again, acrylic yarn is a favorite because it is reasonably priced and comes in a wide variety of colors. However, if your project is something extra special, you may want to spend a little more on tapestry yarn or Persian wool yarn to get certain shades of color.

The types of yarns available are endless, and each grouping of yarn has its own characteristics and uses. The following is a brief description of some common yarns used for plastic canvas.

Worsted Weight Yarn - This yarn may be found in acrylic, wool, wool blends, and a variety of other fiber contents. Worsted weight yarn is the most popular yarn used for 7 mesh plastic canvas because one strand covers the canvas very well. This yarn is inexpensive and comes in a wide range of colors. Worsted weight yarn has four plies which are twisted together to form one strand. When the instructions call for 2-ply or 1-ply yarn, you will need to separate a strand of yarn into its four plies and use only the number of plies indicated in the instructions.

Sport Weight Yarn - This yarn has four thin plies which are twisted together to form one strand. Like worsted weight yarn, sport weight yarn comes in a variety of fiber contents. The color selection in sport weight yarn is more limited than in other types of yarns. You may want to use a double strand of sport weight yarn for better coverage of your 7 mesh canvas. When you plan on doubling your yarn, remember to double the yardage called for in the instructions, too. Sport weight yarn works nicely for 10 mesh canvas.

Tapestry Yarn - This is a thin wool yarn. Because tapestry yarn is available in a wider variety of colors than other yarns, it may be used when several shades of the same color are desired. For example, if you need five shades of pink to stitch a flower, you may choose tapestry yarn for a better blending of colors. Tapestry yarn is ideal for working on 10 mesh canvas. However, it is a more expensive yarn and requires two strands to cover 7 mesh canvas. Projects made with tapestry yarn cannot be washed.

Persian Wool - This is a wool yarn which is made up of three loosely twisted plies. The plies should be separated and realigned before you thread your needle. Like tapestry yarn, Persian yarn has more shades of each color from which to choose. It also has a nap similar to the nap of velvet. To determine the direction of the nap, run the yarn through your fingers. When you rub "with the nap," the yarn is smooth; but when you rub "against the nap," the yarn is rough. For smoother and prettier stitches on your project, stitching should be done "with the nap." The yarn fibers will stand out when stitching is done "against the nap." Because of the wool content, you cannot wash projects made with Persian yarn.

Pearl Cotton - Sometimes #3 pearl cotton is used on plastic canvas to give it a dressy, lacy look. It is not meant to cover 7 mesh canvas completely but to enhance it. Pearl cotton works well on 10 mesh canvas when you want your needlework to have a satiny sheen. If you cannot locate #3 pearl cotton in your area, you can substitute with twelve strands of embroidery floss.

Embroidery Floss - Occasionally, embroidery floss is used to add small details such as eyes or mouths on 7 mesh canvas. Twelve strands of embroidery floss are recommended for covering 10 mesh canvas. Use six strands to cover 14 mesh canvas.

COLORS
Choosing colors can be fun, but sometimes a little difficult. Your project will tell you what yarn colors you will need. When you begin searching for the recommended colors, you may be slightly overwhelmed by the different shades of each color. Here are a few guidelines to consider when choosing your colors.

Try not to mix very bright colors with dull colors. For example, if you're stitching a project using country colors, don't use a bright Christmas red with country blues and greens. Instead, use a maroon or country red. Likewise, if you are stitching a bright magnet, don't use country blue with bright red, yellow, and green.

Some projects require several shades of a color, such as shades of red for a Santa. Be sure your shades blend well together.

Sometimes you may have trouble finding three or four shades of a color. If you think your project warrants the extra expense, you can usually find several shades of a color available in tapestry yarn or Persian wool yarn.

Remember, you don't have to use the colors suggested in the color key. If you find a project that you really like, but your house is decorated in a different color, change the colors in the project to match your decor!

AMOUNTS
A handy way of estimating yardage is to make a yarn yardage estimator. Cut a one yard piece of yarn for each different stitch used in your project. For each stitch, work as many stitches as you can with the one yard length of yarn.

To use your yarn yardage estimator, count the number of stitches you were able to make, say 72 Tent Stitches. Now look at the chart for the project you want to make. Estimate the number of ecru Tent Stitches on the chart, say 150. Now divide the estimated number of ecru stitches by the actual number stitched with a yard of yarn. One hundred fifty divided by 72 is approximately two. So you will need about two yards of ecru yarn to make your project. Repeat this for all stitches and yarn colors. To allow for repairs and practice stitches, purchase extra yardage of each color. If you have yarn left over, remember that scraps of yarn are perfect for small projects such as magnets or when you need just a few inches of a particular color for another project.

In addition to purchasing an adequate amount of each color of yarn, it is also important to buy all of the yarn you need to complete your project at the same time. Yarn often varies in the amount of dye used to color the yarn. Although the variation may be slight when yarns from two different dye lots are held together, the variation is usually very apparent on a stitched piece.

SELECTING NEEDLES
TYPES OF NEEDLES
Stitching on plastic canvas should be done with a blunt needle called a tapestry needle. Tapestry needles are sized by numbers; the higher the number, the smaller the needle. The correct size needle to use depends on the canvas mesh size and the yarn thickness. The needle should be small enough to allow the threaded needle to pass through the canvas holes easily, without disturbing canvas threads. The eye of the needle should be large enough to allow yarn to be threaded easily. If the eye is too small, the yarn will wear thin and may break. You will find the recommended needle size listed in the supply section of each project.

WORKING WITH PLASTIC CANVAS
Throughout this leaflet, the lines of the canvas will be referred to as threads. However, they are not actually "threads" since the canvas is nonwoven. To cut plastic canvas pieces accurately, count **threads** (not **holes**) as shown in **Fig. 1**.

Fig. 1

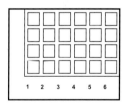

PREPARING AND CUTTING CANVAS

Before cutting out your pieces, notice the thread count of each piece on your chart. The thread count is usually located above the piece on the chart. The thread count tells you the number of threads in the width and the height of the canvas piece. Follow the thread count to cut out a rectangle the specified size. Remember to count **threads**, not **holes**. If you accidentally count holes, your piece is going to be the wrong size. Follow the chart to trim the rectangle into the desired shape.

You may want to mark the outline of the piece on your canvas before cutting it out. Use a China marker, grease pencil, or fine point permanent marker to draw the outline of your shape on the canvas. Before you begin stitching, be sure to remove all markings with a dry tissue. Any remaining markings are likely to rub off on your yarn as you stitch.

If there is room around your chart, it may be helpful to use a ruler and pencil to extend the grid lines of the chart to form a rectangle (see Sample Chart).

Sample Chart

Chicken (18 x 18 threads)

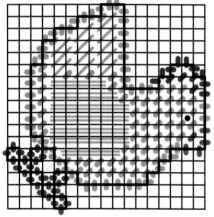

A good pair of household scissors is recommended for cutting plastic canvas. However, a craft knife is helpful when cutting a small area from the center of a larger piece of canvas. When using a craft knife, be sure to protect the table below your canvas. A layer of cardboard or a magazine should provide enough padding to protect your table.

When cutting canvas, be sure to cut as close to the thread as possible without cutting into the thread. If you don't cut close enough, "nubs" or "pickets" will be left on the edge of your canvas. Be sure to cut off all nubs from the canvas before you begin to stitch, because nubs will snag the yarn and are difficult to cover.

When making a diagonal cut on plastic canvas, cut through the center of each intersection. This will leave enough plastic canvas on both sides of the cut so that both pieces of canvas may be used. Diagonal corners will also snag yarn less and be easier to cover.

The charts may show slits in the plastic canvas **(Fig. 2)**. To make slits, use a craft knife to cut exactly through the center of an intersection of plastic canvas threads **(Fig. 3)**. Repeat for number of intersections needed. When working piece, be careful not to carry yarn across slits.

Fig. 2

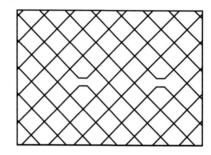

Fig. 3

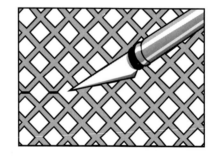

If your project has several pieces, you may want to cut them all out before you begin stitching. Keep your cut pieces in a sealable plastic bag to prevent loss.

THREADING YOUR NEEDLE

Many people wonder, "What is the best way to thread my needle?" Here are a couple of methods. Practice each one with a scrap of yarn to determine what works best for you. There are also several yarn-size needle threaders available at your local craft store.

FOLD METHOD

First, sharply fold the end of yarn over your needle; then remove needle. Keeping the fold sharp, push the needle onto the yarn **(Fig. 4)**.

Fig. 4

THREAD METHOD

Fold a 5" piece of sewing thread in half, forming a loop. Insert loop of thread through the eye of your needle **(Fig. 5)**. Insert yarn through the loop and pull the thread back through your needle, pulling yarn through at the same time.

Fig. 5

WASHING INSTRUCTIONS

If you used washable yarn for all of your stitches, you may hand wash plastic canvas projects in warm water with a mild soap. Do not rub or scrub stitches; this will cause the yarn to fuzz. Allow your stitched piece to air dry. Do not put stitched pieces in a clothes dryer. The plastic canvas could melt in the heat of a dryer. Do not dry clean your plastic canvas. The chemicals used in dry cleaning could dissolve the plastic canvas. When your piece is dry, you may need to trim the fuzz from your project with a small pair of sharp scissors.

GENERAL INFORMATION

1. **Fig. 1, page 93**, shows how to count threads accurately. Follow charts to cut out plastic canvas pieces.

2. Backstitches used for detail **(Fig. 8)**, French Knots **(Fig. 12)**, and Lazy Daisy Stitches **(Fig. 15, page 96)** are worked over completed stitches.

3. Unless otherwise indicated, Overcast Stitches **(Fig. 17, page 96)** are used to cover the edges of pieces and to join pieces.

STITCH DIAGRAMS

Unless otherwise indicated, bring threaded needle up at 1 and all odd numbers and down at 2 and all even numbers.

ALICIA LACE

This series of stitches is worked in diagonal rows and forms a lacy pattern. Follow **Fig. 6** and work in one direction to cover every other diagonal row of intersections. Then work in the other direction **(Fig. 7)** to cover the remaining intersections.

Fig. 6

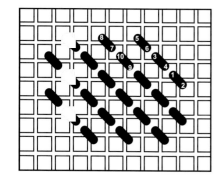

Fig. 7

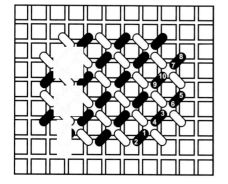

BACKSTITCH

This stitch is worked over completed stitches to outline or define **(Fig. 8)**. It is sometimes worked over more than one thread. Backstitch may also be used to cover canvas as shown in **Fig. 9**.

Fig. 8

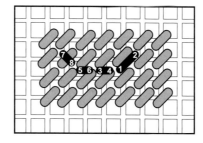

Fig. 9

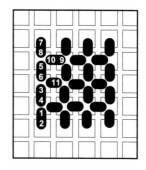

CROSS STITCH

This stitch is composed of two stitches **(Fig. 10)**. The top stitch of each cross must always be made in the same direction. The number of intersections may vary according to the chart.

Fig. 10

DOUBLE FRENCH KNOT

Bring needle up through hole. Wrap yarn twice around needle and insert needle in same hole, holding end of yarn with non-stitching fingers **(Fig. 11)**. Tighten knot; then pull needle through canvas, holding yarn until it must be released.

Fig. 11

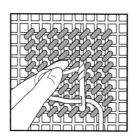

FRENCH KNOT

Bring needle up through hole. Wrap yarn once around needle and insert needle in same hole or adjacent hole, holding end of yarn with non-stitching fingers **(Fig. 12)**. Tighten knot; then pull needle through canvas, holding yarn until it must be released.

Fig. 12

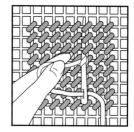

FRINGE

Fold a 12" length of yarn in half. Thread needle with loose ends of yarn. Take needle down at 1, leaving a 1" loop on top of the canvas. Come up at 2, bring needle through loop, and pull tightly **(Fig. 13)**.

Fig. 13

GOBELIN STITCH

This basic straight stitch is worked over two or more threads or intersections. The number of threads or intersections may vary according to the chart **(Fig. 14)**.

Fig. 14

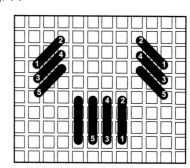

LAZY DAISY

Bring needle up at 1, make a loop and go down at 1 again **(Fig. 15)**. Come up at 2, keeping yarn below needle's point. Pull needle through and secure loop by bringing yarn over loop and going down at 2.

Fig. 15

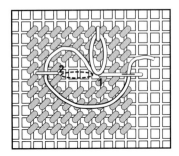

MOSAIC STITCH

This three stitch pattern forms small squares **(Fig. 16)**.

Fig. 16

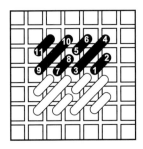

OVERCAST STITCH

This stitch covers the edge of the canvas and joins pieces of canvas **(Fig. 17)**. It may be necessary to go through the same hole more than once to get an even coverage on the edge, especially at the corners.

Fig. 17

SCOTCH STITCH VARIATION

This stitch forms a square with a dimple in the center **(Fig. 18)**. It may be worked over four or more horizontal threads by four or more vertical threads. The Scotch Stitch Variation may slant in the opposite direction.

Fig. 18

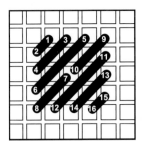

SMOOTH SPIDER WEB

This decorative stitch is made by first making 8 spokes **(Fig. 19)**. Using a one yard length of yarn, come up through the hole under spoke 1-2 and weave over two spokes and back under one spoke until all spokes are covered **(Fig. 20)**. Secure yarn on wrong side.

Fig. 19

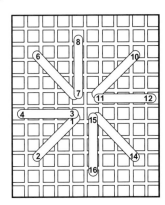

Fig. 20

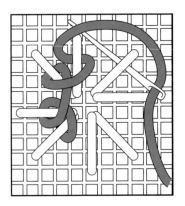

TENT STITCH

This stitch is worked in vertical or horizontal rows over one intersection as shown in **Fig. 21**. Follow **Fig. 22** to work the **Reversed Tent Stitch**.

Fig. 21

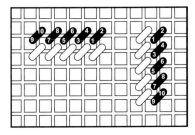

Fig. 22

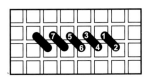

TURKEY LOOP STITCH

This stitch is composed of locked loops. Bring needle up through hole and back down through same hole, forming a loop on top of the canvas. A locking stitch is then made across the thread directly below or to either side of loop as shown in **Fig. 23**.

Fig. 23

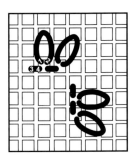

Instructions tested and photography items made by Kandi Ashford, Kathleen Boyd, Virginia Cates, JoAnn Forrest, Janice Gordon, and Vivian Heath.